NINE AND
COUNTING

THE WOMEN OF THE SENATE

2001

Barbara Mikulski

Kay Bailey Hutchison

Dianne Feinstein

Barbara Boxer

Patty Murray

Olympia Snowe

Susan Collins

Mary Landrieu

Blanche L. Lincoln

Debbie Stabenow

Maria Cantwell

Hillary Rodham Clinton

Jean Carnahan

NINE AND COUNTING

WRITTEN WITH

CATHERINE WHITNEY

Perennial

An Imprint of HarperCollins*Publishers*

To the Girl Scout Organization,
which, for eighty-nine years, has helped
young women realize their full potential;
and
to the young women of America,
who can create whatever you can dream.

CONTENTS

ACKNOWLEDGMENTS

Like the work we do every day in the United States Senate, writing this book has required the commitment and cooperation of many people. We are extremely grateful for the shared effort that has made it possible for us to communicate to other women the vast possibilities that exist for all of those who are called to public service.

We are especially appreciative to our editor, Claire Wachtel, for her vision and guidance. She shepherded the book from the beginning, with enthusiasm and skill, and made everything possible.

Catherine Whitney, our collaborator, gave voice to our ideas, beliefs, and experiences, effectively balancing the needs of nine very different women. We also appreciate the contributions of Jane Dystel, Catherine's literary agent, and Paul Krafin, who assisted with the development of the manuscript.

Washington lawyer Bob Barnett and his associate, Kathleen Ryan, were invaluable in crafting the project in its early stages, and giving the best advice at every turn.

Annie Leibovitz generously donated her time and tremendous talent to shoot the wonderful photograph on the cover. The setting for this photo was the Old Supreme Court Chamber, a stunning historic site, whose choice was made more significant by the fact that women were not allowed on its floor in the days of its active use, between 1810 and 1860.

We are also thankful to Jason Schmidt, for allowing the use of his wonderful photograph of all thirteen women senators on the back cover, and to Tom Booth and his agency for expediting the process.

Our staffs make it possible for us to be effective on every level. Their special efforts on behalf of this book are greatly appreciated. In particular, we'd like to mention Larry DiRita, former chief of staff to Senator Kay Bailey Hutchison, and Jenny Luray, chief of staff to Senator Barbara Mikulski, who served as the guardians of the hardcover and paperback editions. We'd also like to give special thanks to Tina Giordano Henry, Senator Hutchison's scheduler, who tirelessly and cheerfully worked to coordinate the senators' schedules for meetings and interviews.

In addition, we are grateful for the efforts of other staff members who helped to make this book a reality: Johanna Ramos-Boyer, press secretary for Senator Mikulski; Howard Gantman, communications director for Senator Dianne Feinstein; the staff of Senator Barbara Boxer; Felicia Knight, press secretary, Kimberly Hamlin, former deputy press secretary, and Cynthia Bailey, appointment secretary, for Senator Susan Collins; Gina Farrell, deputy communications director for Senator Mary Landrieu; Drew Goesl, press secretary, Jennifer Martinez, deputy press secretary, and Jennifer Greeson, former press secretary, for Senator Blanche Lincoln; Todd Webster, communications director, and Tovah Ravitz, former press secretary, for Senator Patty Murray; Kevin Raye, chief of staff, Dave Lackey, communications director, Jane Calderwood, legislative director, and John Richter, special assistant, for Senator Olympia Snowe; Mary Beth Schultheis, press secretary for Senator Maria Cantwell; Kerin Polla, communications director, and Jean Marie Neal, chief of staff, for Senator Debbie Stabenow; Karen Dunn, communications director, Matthew Nelson, and Heather King, for Senator Hillary Rodham Clinton; and Roy Temple, chief of staff, Tony Wyche, communications director, and Melissa Schwartz, scheduler, for Senator Jean Carnahan.

We appreciate the contributions of many talented and dedicated people at William Morrow/HarperCollins: Jennifer Pooley, assistant to Claire Wachtel; Laura Leonard, Dominique D'Anna, and Cathy Saypol.

Special appreciation to Senator Robert C. Byrd, who has contributed a substantial body of work to the Senate's historical archive. Thanks to his efforts, the stories of the women who paved our way have been preserved.

Finally, we are thankful to all of the women who have served before us in the United States Senate. Their pioneering spirits and determination opened the door for us so that now we can open that door for others.

The proceeds from the sale of Nine and Counting *are being donated to the Girl Scout Organization.*

INTRODUCTION

The Power of Nine Women . . .
Just Like You

by Catherine Whitney

When I was a young girl, my mother was involved with the local political scene in Seattle, Washington. She used to take me along when she went doorbelling for local candidates. We'd cover her assigned territory, Forty-third and Forty-fourth streets, on both sides, stopping at each house to deliver a flyer and talk to a neighbor. Sometimes we'd go over to a candidate's home and spend an evening stuffing envelopes. I can remember sitting at long tables with other kids and their moms, engaged in an assembly-line process of folding, stuffing, and licking. By the time we finished, our hands would be stained blue from the mimeographed sheets of paper, and our mouths would be parched from the sticky glue of the envelopes.

Ringing doorbells and stuffing envelopes were jobs mostly left to women. Usually, the only men who made appearances were the candidates themselves. In those days—the 1950s and early sixties—I never saw a woman candidate, but almost all of the campaign workers were women.

At school, in our fifth-grade civics class, we learned about the nation's only woman senator, Margaret Chase Smith of Maine. Our teacher spoke reverently of her magnificent achievement, and she seemed far removed from our everyday reality. Margaret Chase Smith was placed on a pedestal along with other female icons—Florence Nightingale, Amelia Earhart, and Eleanor Roosevelt. We admired them, but were not expected to emulate them.

My parents, Dick and Janet Schuler, were ordinary people who believed in government as a tangible entity that had an impact on their lives. Politics was a regular topic of conversation, sometimes producing fireworks when they didn't agree on an issue or a candidate. On several occasions during the 1960 presidential campaign, my parents waged their own heated, and much louder, version of the Kennedy-Nixon debates at our dinner table.

Although it was rare for a woman to run for or be elected to public office during those years, we were all raised to be civically involved. Participation in the political process was both a privilege and a duty. Ringing doorbells, stuffing envelopes, debating the merits of candidates, voting—these were activities to be cherished.

When my Girl Scout troop took its annual trip to the state capitol in Olympia, Washington, the polished hallways, high ceilings, and wide staircases were like echo chambers of American history. They never failed to impress us, and our voices fell into hushed whispers as we took our guided tour.

Today, when the Girl Scout troops visit the state capitol, they may feel the same sense of awe. They will also witness something we never did—women working alongside men in the legislature. Their own representative might be a woman. Today, young girls can participate in campaign activities, as I once did, but the candidates they support may likely be women. They might even be their own mothers. Washington State currently has a higher number of women legislators than any other state. Civics teachers can now clearly trace the progress of women in the political arena. They can point to thirteen women in the United States Senate, two of them from their own state. They can talk about sixty-one women in the U.S. House of Representatives, and five women governors. In recent years, women have filled important positions in the president's cabinet, including those of attorney general, secretary of state, and director of the EPA. There are two women on the Supreme Court. These women are no longer presented as elevated and unreachable icons, but as true role models for others to emulate and follow.

When you are raised to believe in government and to love the mechanics of democracy, those values remain at the core and never leave you. They transcend political parties, stay alive through disappointment and scandal, and keep you energized for each new campaign. This has been my mother's legacy to me, and I am pleased to be working with the nine women of the United States Senate to pass that legacy on to young women today.

Nine and Counting will acquaint you with nine real women who have faced and overcome great challenges to reach the United States Senate. This edition also introduces the four women who were elected to the Senate in 2000. By telling of their experiences, they hope to serve as guides to other women who face similar challenges. Because they have succeeded, others can succeed, too.

The women of the United States Senate offer a picture of optimism and opportunity for women everywhere. By their courage, their stamina, and their actions, each of them has proved that public service is precisely that—an opportunity to serve the greater good, a chance to make a measurable impact on the world around them. They also understand that women are urgently needed to bring their special perspective to the public debate. As the author Carol Gilligan states so eloquently: "In the different voice of women lies the truth of an ethic of care, the tie between relationship and responsibility, and the origins of aggression in the failure of connection."

This book has been written as an invitation by the women of the United States Senate to every young woman in America. They ask you to join them and let your voice be heard.

NINE AND COUNTING

PROLOGUE

The Dinner Club

February 2000—The Monocle

Tucked away on a quiet street a block from the Senate offices sits The Monocle, the quintessential Capitol Hill restaurant. It is housed in a modest little building with an auspicious history. Enter its doors, look around, and you are immediately transported back to another era. The interior has all the trappings of an exclusive men's club—the faded maroon carpeting, the flocked red-and-gold walls, the discreet banquettes set off from the open dining area. Gold-leafed political quotes are scrolled along the beams that span the circumference of the room:

It's lonely up here.

Government is not an exact science.

I give special consideration to everybody.

The Monocle opened its doors in 1960, during the Kennedy-Nixon campaign. It still carries the faint whiff of a time when powerful men brokered deals over steaming bowls of clam chowder and heaping platters of seafood, and coaxed the reluctant over to their side with brandy and cigars. Networking took place in a traditional way. It is the method of the powerful at ease with one another—old friends smoothing the path, trying to deal with the gritty business of government.

The crab cakes at The Monocle are still ranked among the best in D.C., but the men's club has faded into history. All that remains are the framed photographs that line the walls—Lyndon Johnson, Barry Goldwater, Jack Kennedy, Tip O'Neill, Warren Magnuson, Richard Nixon—the old guard who have long since passed away.

Tonight, up a red-carpeted flight of stairs, in a private dining room, the table is set for nine. Three uniformed waiters silently arrange chairs and fill water glasses beneath the gaze of a dozen portraits of past Supreme Court justices. Nine of the most powerful people in America—the women of the United States Senate—will be having dinner here.

For several years now, the Senate women have met for dinner every few weeks. Sometimes they'll go to The Monocle, because of its proximity to their offices. On other occasions, they'll dine at a senator's home or try out a new Washington restaurant. These ad hoc dinners are informal gatherings. Still, as the women take their places around the table this evening, one imagines that the old-timers watching from the walls would undoubtedly be unnerved by the sight. Perhaps they'd wonder what plots are being hatched here over salads and crab cakes; what deals are being made over white wine spritzers and diet sodas.

The purpose of these regular dinners is neither plot hatching nor deal making. It is, rather, a familiar ritual among women colleagues everywhere—that uniquely female manner of lending support by sharing experiences, describing challenges, and talking about the issues they care about. For a couple of hours every month, they are able to relax, away from the ever-present staffs, the ringing phones, and the demands of public life.

In the ordinary course of their days, they don't constantly think of themselves as "women senators." They are just senators—individuals within a group of one hundred, doing the work of the People. While they sometimes unite behind a piece of legislation that has special implications for women, they do not have their own caucus. Indeed, their positions on most of the issues run across the spectrum, from conservative to liberal and those in between. Each senator has her own agenda for what she wants to accomplish in office. There is no singular *women's* agenda.

These nine women who are most distinguished by their differences come together not because of what they believe, but because of who they are.

Gender has been the transcendent characteristic of their personal and professional lives. They were raised in an era when women were not encouraged to seek political office—yet here they are. They have reached the Senate by overcoming enormous obstacles. All of them have encountered resistance to their dreams and ideas, not because they lacked brainpower, ambition, ability, or charisma, but because they were women. They have been forced to make choices not often required of male candidates— to invent the roles of politician as wife, mother, daughter, sister, grandmother. Perhaps, too, their successes have been sweeter because they have been gained at such a high cost.

These common experiences enforce a bond that might not otherwise exist in such a mixed group. The senators *know* each other on a fundamental level.

And so they dine and talk, as all women colleagues do in such informal settings—about crime on the streets, bottlenecks on the floor, the death penalty, and the price of gas. They argue the merits of a piece of legislation, as well as the merits of a popular movie. They recount funny stories. They talk shop, but they don't gossip.

In about two hours, after the decaf coffee has been served and the temptations of the dessert menu refused, they begin to leave, carrying the bulging briefcases that contain their night reading.

As they walk down the stairs, a more senior senator murmurs sympathetically to one of the others, "You look tired. Are you okay?"

She receives a faint smile. "Yes. Thanks for asking. It's been a long day."

Tomorrow they'll be back in their separate universes, fighting for their different goals. But for a couple of hours tonight, they have participated in a simple, energizing ritual of unity.

1

Why Not a Woman?

In America any boy may become President.

ADLAI STEVENSON, September 1952
Presidential Campaign Speech

In 1952, when Adlai Stevenson was rhetorically passing the torch of democracy to every boy in America, he didn't think to mention what "any girl" might become. But for nine women, that question was already part of their destiny.

In 1952, Barbara Mikulski was sixteen years old. At the Catholic high school she attended in a Polish enclave of Baltimore, she joined the Christopher movement, which promoted service to the poor and comfort to the suffering. Even then, she sensed that she had a rendezvous with destiny—that through her work she might light a candle that would cast a wide beam. The desire to help others had been cultivated in Barbara from an early age. Raised in a hardworking, close-knit community of first- and second-generation Polish families, she saw daily examples of the ways in which small acts of charity could transform lives. Barbara knew that she wanted to dedicate her life to service. She imagined that such a choice would lead to work in the social services or health care. At sixteen, she didn't make the connection between her aspirations and work in government, and even if she had, she might have dismissed the notion that government could provide such an opportunity for a woman. Barbara Mikulski's place in the United States Senate would be earned, thirty-four years later, through her labors in the community, and she would carry the lessons of the grass roots with her into the corridors of power.

Kay Bailey was nine years old in 1952, living an unexceptional middle-class life in La Marque, Texas, a small town near Galveston. She was a busy, gregarious little girl, forever organizing skits and projects among the neighborhood children, and participating in activities with her Girl Scout troop. On Saturdays, her mother drove her to Houston, fifty miles away, where she studied ballet and performed with the Houston Youth Symphony Ballet. She had boundless energy. Kay's dreams of the future took her from La Marque to the exciting urban arenas of Houston and Dallas, where she aspired to a profession in business or the law. Washington, D.C., was far beyond the scope of her dreams, and politics further still. Her family wasn't particularly political. Yet by the time Kay Bailey Hutchison was elected as the first woman from Texas to serve in the United States Senate forty-one years later, she would have made her mark on the political landscape of the state.

In 1952, as nineteen-year-old Dianne Goldman began her

studies at Stanford University, she already possessed a keen interest in politics. It had been planted early on and nurtured throughout her young life by her father's brother Morrie. Uncle Morrie, a colorful and vociferous coat manufacturer, introduced his niece to the ins and outs of city hall, and engaged her father in heated debates at the dinner table. While Morrie sparked Dianne's imagination, her father, Leon, was her true mentor. A prominent San Francisco surgeon, Leon instilled in his daughter the fundamental principles that would remain with her for life. Her admiration for her father was so great that she briefly considered a career in medicine. However, an A-plus in American political thought and a D in genetics during her freshman year in college convinced her to follow her passion—politics. Dianne Goldman was raised to believe that she could accomplish anything she set out to do. Dianne Feinstein would confirm that belief repeatedly throughout a lifetime of public service, culminating with her election in 1992 to represent the state of California in the United States Senate.

In 1952, Barbara Levy was a spirited ten-year-old, growing up in a nice middle-class section of Brooklyn, New York. Extroverted and irrepressible, she was a popular student—"perky, peppy, happy," in her own words—the girl who organized pep rallies and got chosen "all-around camper" in the summer. Barbara had a bright mind and an innate curiosity. She especially enjoyed political science and economics in school. But these interests were asides to her true ambition, which perfectly reflected the cultural ideal of the early 1950s. It was widely accepted that women, once they completed their education, would devote themselves to raising their children. Barbara's mother often spoke sympathetically of women who "had to" work, and Barbara assumed that her life would follow a predictable path: she would continue her schooling and get a good education, fall in love, marry, have children, and live the American dream. All of that did happen, but the path was not at all predictable. By the time Barbara Boxer claimed victory in her race for the Senate forty years later, she would have charted a course that would have seemed unthinkable to her ten-year-old self.

Five-year-old Olympia Bouchles was starting kindergarten at the Wallace School in Lewiston, Maine, in September 1952. A serious and responsible little girl, she was the daughter of a Greek immigrant father and a first-generation Greek-American mother. When Olympia came home from school at noon, her father, who was determined his daughter would have a strong and early start on education, sent her back each day saying, "How do you expect to learn if you come home at noon?" Olympia had to ask her teacher to send a note home to explain that she wasn't skipping school; kindergarten was only a half-day.

Olympia's hardworking mother and father had died by the time she was a fourth-grader, but their early influence stayed with her as she expanded her horizons to the world beyond her hometown. She grew up, fulfilled her parents' dream by graduating from college, and married. She was proud when her husband, Peter Snowe, won election to the Maine Legislature. But after less than three and a half years of marriage, and just three months into his term, he died in an automobile accident. Twenty-six-year-old Olympia summoned the inner resources instilled in her by her parents and her aunt and uncle and turned her grief into action. She ran for her late husband's seat and began a political career that would see her become the only woman in the history of the nation to be elected to both houses of a state legislature and both houses of the United States Congress.

In 1952, Patty Johns was an energetic two-year-old, one of a set of twins born in Seattle, Washington, and growing up in the booming postwar economy of Bothell. She and her twin sister, Peggy, were two of seven children whose hardworking parents raised them with strong traditional values. The Johns children all worked in the five-and-dime store their father managed on Maine Street. When Patty was fifteen, her father became debilitated by multiple sclerosis and was forced to retire. Everyone helped out, and Patty and her siblings all worked to put themselves through college. Her family instilled in Patty a sense that by working together they could solve problems that could not be solved individually. Today Patty Murray brings that philosophy to the United States Senate, a platform she never could have dreamed of then.

In September 1952, Susan Collins was three months away from being born as the third of what would be six children to Don and Pat Collins in Caribou, Maine. The Collins family roots run deep in Caribou, a northern Maine town that is equidistant from the state capital and Quebec City. Susan's ancestors founded the town in the 1840s, and a few years later her great-great-grandfather started the S. W. Collins Company, a small lumber business that today is run by Susan's younger brothers, Sam and Gregg. Small business is not the only calling to which the Collinses have traditionally been dedicated; five generations of Collinses, including Susan's father, served in the

Maine State Legislature. Her mother, too, was active in the community, serving as chairman of the Caribou School Board the year Susan graduated from high school. Susan grew up hearing the call of public service and knowing that she, too, wanted to give something back to her community. When Susan was sworn into the United States Senate in 1997, the first bill she introduced was the Family Business and Farm Preservation Act, a bill that sought to make it easier for family businesses to be handed down from generation to generation.

In 1952, the year that Dwight D. Eisenhower was elected to his first term as president of the United States, neither Mary Landrieu nor Blanche Lambert was yet a gleam in her parents' eyes. Mary would be born three years later in Virginia, while her father, Moon Landrieu, the future mayor of New Orleans, was serving a minor stint at the Pentagon. The eldest of nine children in a large, loving New Orleans family, Mary was taught the value of responsible leadership from an early age. Exposed to her father's career, she was comfortable in political circles, but her dream was not politics, but social work. That changed early on, and at age twenty-three, Mary was first elected to the Louisiana legislature. Eighteen years later, she was elected to represent Louisiana in the United States Senate.

Blanche Lambert would arrive in the world in 1960. Her birth, in Helena, Arkansas, as the youngest of four children, would precipitously occur at the turn of a decade that has come to symbolize a monumental cultural shift in America. In the Lambert home, the bedrock values still prevailed. The Lamberts were farmers, and they cultivated in their children a deep love for the Arkansas land and a strong sense of ownership in the rights and responsibilities of their citizenship. Wanting to make a mark on the world, Blanche set her sights on a career in the medical sciences; her college major was biology. After college, she took a job in Washington, D.C., and fell in love with government. She went on to serve in Congress, representing the people of Arkansas. In 1998, Blanche Lambert Lincoln became the youngest woman ever elected to the Senate.

When Adlai Stevenson promised the boys of America the brass ring of democracy, that promise was a faint, buried hope seldom articulated by girls. If progress is a weighted continuum, with each progressive act adding its impact to the next, then the paths of America's young girls were inexorably fated to converge with an era of unlimited possibility. But it wouldn't happen overnight.

In the eighty years since women won the right to vote, the glass ceiling that stifled their political ambitions has cracked, but has yet to break. At times the progress has been so slow as to seem nonexistent. Very occasionally, a breakthrough has sent a large shard crashing down. As Dianne Feinstein remarked, eight years after her election to the Senate, "The glass ceiling is chipped. It's not yet shattered."

The United States Senate has always been viewed as one of the greatest of political prizes—a setting where individuals have the power to influence our greater national goals, as well as provide a direct and important source of support at the state and local levels. With only one hundred members, the Senate is also a more intimate setting for discussion, understanding, and amelioration of the issues that face the nation. It has been said that the United States Senate is the most exclusive club in the world. For most of our history, it has been the most exclusive *men's* club. No more. Since 1922, twenty-seven women have served in the Senate—

if one uses the term "served" loosely. In fact, from the moment in 1922 that Rebecca Latimer Felton of Georgia took the oath of office to become the first woman senator, until 1992, when the election raised the number of women senators to six, only five women served terms lasting more than a few months. Their terms were mostly symbolic. Felton was perhaps the most symbolic of all, serving a term of office that lasted precisely one day.

October 1922—Washington, D.C.

Georgia governor Thomas Hardwick must have thought it was a stroke of political genius when he appointed Rebecca Felton to finish out the Senate term of Thomas E. Watson, who had died suddenly while in office. Since Hardwick was running for the Senate seat himself, he had a personal interest in appointing someone who would not present a challenge to his own ambitions. An eighty-seven-year-old woman could hardly be considered a threat. At the same time, Felton was a woman of substantial achievement, highly respected among the newly enfranchised women voters. Hardwick was interested in courting Georgia's female voters, and he needed a bold action to distract them from the memory of his fierce opposition to the passage of the Nineteenth Amendment two years earlier. He had much to gain and nothing to lose by appointing Felton. It wasn't as if she would actually serve. The Senate was out of session at the time, and it would not reconvene until after the election. Hardwick fully expected to be the newly elected senator by then, and it wouldn't even be necessary for Felton to make the long trip from Georgia to take the oath of office.

It would have been a perfect plan, but for one thing. Thomas Hardwick lost the election to Walter George. Perhaps the women of Georgia were not so easily distracted, after all.

Before the Senate convened in November 1922, Senator-elect George made a historic gesture. The gentlemanly Southerner announced that he would step aside for one day so that Rebecca Felton could take the oath of office that would secure her place in history as the first woman senator.

A large crowd of women packed the Senate galleries on November 20 to watch the historic proceedings. Walter George's gesture may have been well meant, but it's doubtful that he fully understood the power of the moment, or the impact it would have on the crowd when Rebecca Felton walked in.

Indeed, Rebecca Felton had a well-established reputation as an outspoken activist, whose caustic wit and acerbic statements had grown more piercing with her advancing years. She had traded in her opportunity to be a Southern belle long before, instead embracing a lifetime of passionate causes. She'd long been at the forefront of the suffragist movement, and had lectured extensively on matters ranging from agriculture to the military to the oppression of slaves. And so when Rebecca Felton strode slowly down the aisle to take her place, the gallery erupted with ecstatic cheers. Felton paused in mid-stride, turned in the direction of the gallery, and blew her supporters a kiss, sending them into fresh waves of cheering.

But before Felton could take the oath of office, the Senate adjourned after just twelve minutes—out of respect for their deceased fellow senator Tom Watson, whose seat Felton was about to assume.

Undaunted, Rebecca Felton and her gallery of supporters returned the next day, November 21, and this time, no one went away disappointed. Felton's remarks before the Senate, in the brief time she was officially a member, sent a strong, clear message. Standing proudly as the "junior senator from Georgia," she faced the men who would be her colleagues for the day. She began by describing a cartoon she had recently received, picturing the Senate in session. "The seats seemed to be fully occupied," she said, "and there appeared in the picture the figure of a woman who had evidently entered without sending in her card."

Felton cast her powerful gaze at her fellow senators, arching her deeply wrinkled brow to expose the twinkling mix of irony and scorn that lit her eyes. "The gentlemen in the Senate took the situation variously," she continued. "Some seemed to be a little bit hysterical, but most of them occupied their time looking at the ceiling." She pointed out that no one had bothered to offer the lady in the cartoon a seat. Left unsaid was the obvious—Rebecca Felton had not really been offered a seat, either.

Rebecca Felton's concluding remarks were offered both as a promise and as a rebuke: "When the women of the country come in and sit with you, though there may be but a very few in the next few years," she told the senators, "I pledge to you that you will get ability, you will get integrity of purpose, you will get exalted patriotism, and you will get unstinted usefulness."

Although the official list of women who have served in the Senate numbers twenty-seven, many of them reached the chamber as did Rebecca Felton—"without sending in her card." During most of the twentieth century, almost every woman who joined the Senate did so by authority of a death certificate—usually belonging to her husband. Twelve women were appointed or elected to finish the terms of dead senators, and most of them served very short terms of only two to four months. Two of them were never sworn in, as the Senate was not in session during their brief appointments. Hattie Caraway of Arkansas, who followed Rebecca Felton a decade later as the second woman senator, was a notable exception. Although Caraway was initially appointed to fill the vacancy created by the death of her husband, she went on to be reelected twice, and she served in the Senate until 1945. Not until 1948, when Margaret Chase Smith was elected in her own right, would a woman gain the office independent of the death of a man—although Smith's initial election to political office, as a member of Congress, did occur in a special election following the death of her husband, Clyde H. Smith, in 1940.

In terms of true senatorial power—the kind that comes from being elected to serve a full term—only fifteen women have reached that pinnacle, and only six before 1992:

- Hattie Wyatt Caraway, Democrat from Arkansas, 1931–1945
- Margaret Chase Smith, Republican from Maine, 1949–1973
- Maurine Brown Neuberger, Democrat from Oregon, 1960–1967
- Nancy Landon Kassebaum, Republican from Kansas, 1978–1997
- Paula Fickes Hawkins, Republican from Florida, 1981–1987
- Barbara Mikulski, Democrat from Maryland, 1987–present

2

"Take Notice . . . and Remember"

I had found what I was meant to do.

DIANNE FEINSTEIN

November 1999—Beaumont, Texas

Beaumont might just be another small town in southeastern Texas if modern Texas—the Texas of black gold—hadn't erupted from a hole drilled at a local salt dome called Spindletop in January 1901. Beaumont, with a current population of a little over 114,000, and a surrounding population of nearly half a million, became a major oil center. It sits on the banks of the Neches River, and is the seat of surrounding Jefferson County. The discovery of oil, and the subsequent completion of a channel to the Gulf of Mexico in 1916, led to its becoming heavily industrialized. It is also positioned at a geographic and cultural crossroads. Only thirty miles from the border of neighboring Louisiana as well as the Gulf of Mexico, its mingled traditions range from country western to Mexican to Cajun. Dynamic and diverse, Beaumont has seen both prosperity and hardship in its 162-year history.

The L. L. Melton Family YMCA is a worn one-story building, set in a largely African-American working-class neighborhood of Beaumont. The center of social services and youth programs, the Melton Family Y has struggled to keep pace with the growing needs of its community. On this day, the small gymnasium is packed with children, parents, staff, civic leaders, and Texas dignitaries. It is an important occasion, and an unprecedented day in the life of the community.

A current of excitement jolts the crowd as the local congressman, Nick Lampson, is joined by Senator Kay Bailey Hutchison. They've come to formally present a $500,000 federal grant for the renovation and expansion of the center. Congressman Lampson is a Democrat, Senator Hutchison a Republican.

The executive director of the YMCA, Maurice Hill, an intense African-American man, addresses the children, who are seated in a semicircle on the floor in front of the podium.

"I want you kids to take notice and remember," he says with pride. "In Washington, D.C., where people say nothing ever gets done, two people from different political parties have worked together to make possible the largest single gift ever given to the children of Beaumont."

The YMCA grant is a triumph of collaboration—the kind of behind-the-scenes grass-roots endeavor that shows governmental process at its best. Positive results such as this grant rarely make the local news, but reflect the bedrock of the democratic ideal. These are the efforts that reinforce the true meaning of representative government. There is little room for partisanship as political officeholders serve the needs of their constituents. At the local level, the ideals and needs of all citizens meet on the most mundane of playing fields. Senator Hutchison will often say—and it is clear that she means it—that it is this kind of tangible achievement that drives her ambition.

When she rises and moves to the podium to speak, her eyes are only for the children seated around her. Hutchison is always delighted whenever she has an opportunity to talk to children, and there's an unmistakable chemistry between them. They seem to regard her as she does them, with a wondrous awe. Hutchison tells them how much she admires them, and urges them to study hard and stay involved in their community. Before she's finished, she gives each of the children a bookmark as a memento of the occasion.

A young girl runs up to her, waving her bookmark. "Ma'am, will you autograph this for me?" Hutchison takes out her pen and signs her name as the other children crowd around. The senator is soon surrounded by a huddle of chattering, laughing children. They all want their bookmarks autographed. Clearly enjoying herself, Hutchison lingers to sign the bookmarks and chat with the children. Her staff, always conscious of time and the need to move on quickly to the next event, finally pull her away. She leaves the building exhilarated.

"Well, I feel like a rock star!" she exclaims. "Do you think those children will remember?" She immediately answers her own question. "Maybe. I remember the first time I ever saw a senator. He visited my hometown, and my dad took me out of school to see him. I was awed, but I never thought I would succeed him someday. *Never.*"

Kay Bailey Hutchison understands that every time she appears before a group of young people, it will probably be the first time they've ever seen a woman who's a senator. She finds satisfaction in the fact that she can serve as a role model—that young girls might aspire to lives of public service because of her.

Growing up in the small town of La Marque, Texas, Hutchison had no female mentors—role models who could provide direction for her natural gregariousness, her creative spirit, and her drive to achieve. While her mother was always loving and supportive of her daughter's projects, she was plainly baffled by the ball of energy that was her child. Hutchison smiles in recollection. She was so full of ideas, even as a very young girl.

KAY BAILEY HUTCHISON

When I was about ten or eleven, I came up with a plan to run a summer school at my house, like a camp, for the neighborhood kids. My mother just shook her head in amazement. She couldn't understand where I came up with such an idea. But she was supportive. She helped me set it up, and we had a week's summer camp in the back den of our house, and all the little kids in the neighborhood came. I had a program for them every day.

Why did I do it? I don't really know. People might say it was a precursor, but I never perceived myself as being different from the other girls. It wasn't as if I consciously had ambition. I never thought about being a senator, or anything like that.

After college, though, I knew I didn't want to do what all of my friends were doing. They were getting married. Then most of them would be schoolteachers. They'd move to Houston or Dallas, and they'd teach school and raise their families. That was mainly it. I wasn't ready to get married, so I decided to go to law school. I'd become a civil attorney, and work in a good law firm, and someday make partner.

I was one of only five women in my class of five hundred at the University of Texas School of Law. I loved law school. It was the first time I really enjoyed school as school. It was the first time I had ever been intellectually stimulated and challenged. When I graduated in 1969, I couldn't wait to get out and prove myself. I was completely unprepared for what happened next. Even though my male classmates were getting hired by all of the big law firms, I was running into a wall. I interviewed with about thirty firms, and the response was always the same. They would compliment me on having graduated from law school, and tell me they were sure I'd make a fine lawyer. Then they'd give me the speech: "We have to invest so much in starting lawyers, and we would lose money on a woman. Training is expensive, and you will get married and move away, or get married and get pregnant, and we can't afford to put all that money in and not get a return. Sorry . . ."

It was a tremendous blow. Until then, my life had been great. I had always been successful—able to do anything I wanted to do. I had all of the confidence in the world, and suddenly it was meaningless, because I couldn't get a job. I had to face my friends, who were going to work for all the good firms, and I was unemployed. I felt so low. My parents were supportive, as always, but it was my first life crisis.

One day, after yet another disappointing interview, I was driving home and I passed TV station KPRC, the NBC Houston affiliate. On a lark, I pulled into the lot, went inside, and asked the receptionist if I could speak to someone about a job.

"What kind of a job?" she asked.

I said, "A news reporter's job."

"You mean you want to talk to the news director?"

"Sure." That sounded right. I knew nothing about the news business, but I was trying to sound confident.

Remarkably, Ray Miller, the news director, agreed to see me, which, I later learned, was very unusual—especially since there were no women news reporters on any of the Houston television stations. He was intrigued with the idea of a woman with a law degree. He didn't hire me on the spot—it wasn't that easy. But a few weeks later, he called and offered me a job as a reporter. Because of my law degree, he decided to experiment by assigning me to cover the state legislature in Austin. It was the first time a local TV station had set up a state capital bureau. It was also the genesis of my life in government—the start of a career path I could never have predicted.

When I took that job, I didn't think of it as a potential career. It was just a chance to get some experience before moving on to something else. I thought of it as a way station and it turned out to be a heck of a way station. I was on TV every night, and people got to know me. I learned the ins and outs of the legislature. When the Harris County Republican chairman asked me to consider running for the legislature, I was surprised. Then the idea started growing on me. I realized that I wanted to do it.

When I look back on it now, I can see that those rejections from law firms were the beginning of my future, even though at the time I thought they were the end. That crisis, and being able to overcome it, gave me a new strength and maturity. When I speak to groups of young people, I always tell them, "Never give up. If a door closes, open a window."

The promise of democracy is that ordinary people can make an extraordinary difference. Increasingly, it is the women of our nation who have tested that core value. Excluded from the political machines, the men's clubs, and the hierarchies of corporations and law firms that have traditionally fostered each new generation of public officials, women have typically arrived at their positions as outsiders. Perhaps this electoral disadvantage affords them a legislative advantage. They can more readily identify with the concerns of everyday citizens, who often feel like outsiders themselves.

Senator Patty Murray once commented, "Almost every woman I've ever met in politics got into it because she was mad about something." Women have been more likely to join together for organized political action because of issues that directly influence the well-being of their families and communities. Most of the women now serving in the Senate never considered political life until one of those issues triggered a fierce resolve. Government service became a goal and a reality only as an extension of their grass-roots efforts.

Olympia Snowe, for one, always thought she would work in

government. But as she related years later as a congresswoman in her first speech to a graduating class of her alma mater:

OLYMPIA SNOWE

When I graduated from college . . . I found conflicting choices in my life. At the time I was engaged to be married, yet I also had my own aspirations as an individual. I was anxious that with the words "I do" all my goals and dreams would evaporate.

And although politics wasn't high on the menu of choices for women back in 1969, I knew that I wanted to be involved in some form of public service . . . in playing a part to help improve the lives of others. What form that service would take was quite another mystery. Back then I was not even contemplating elective office. Nor did I realize that the meaning of my life's experiences wouldn't really surface for some years to come. That, for example, having only made it through the University of Maine with the use of student loans would teach me their critical value years later in Congress.

Or that working in a Christmas ornament factory during the summer would teach me the enormous importance of workplace safety—again, something that would surface as a significant issue in Congress.

I did know this, however: Whatever course my life would take, I wanted to be relevant to the world around me. And during my time in Congress, I have always tried to apply the lessons I learned earlier in life toward making a difference for others.

It is clear that the nine senators understand that they are seen as role models for an emerging population of young women. For Senator Mary Landrieu of Louisiana, the goal is, quite literally, to provide a picture of possibility.

MARY LANDRIEU

All my life I've seen campaign photos of male politicians surrounded by their wives and children. I think every male politician has a picture like that. I always thought it would be nice to see women in those pictures, surrounded by their husbands and children. I was determined to make that picture possible for my state. I was determined to stay in office *and* have a family and children. I wanted young women to know, if you've been called to lead, you can do it. You don't have to abide by different standards than men.

It breaks my heart when I meet older women who once made a choice between a career and a family. There was a time, not long ago, when many women *had* to make that choice. Now these women are retired, and they have no children, no grandchildren. In some cases, not all, they were forced to sacrifice one great joy for another. It just doesn't seem right. I want to make sure that picture is changed for good. If I can do it, other women can.

There are other issues as well. Women in public service make a point of stressing the unique perspective they bring to the table. It is not their intention to encourage young women to be more like men. The dynamics of power and achievement are different for women, and it is important to preserve those gifts. "Our challenge," California senator Barbara Boxer says, "is to send the right message. How do we encourage young women to pursue careers in public service starting at the grass-roots level, just as we all did? Without the valuable lessons that training ground provides, I'm afraid we'll lose the tradition we've built of a genuine, issues-oriented approach to problems. It isn't necessarily our goal to have young girls all over the country declaring, 'I want to be a senator when I grow up.' We need to encourage them to get involved in the issues that affect them locally first."

Barbara Mikulski was once asked if she'd ever dreamed as a young girl that she would someday achieve a high public office. She laughed at the notion and answered: "I didn't sit around in my little sandbox in a Baltimore ethnic neighborhood saying, 'Oh, one day I'm going to be a U.S. senator.'"

Dreams of reaching high public office were not typical in the East Baltimore neighborhood of Highlandtown. In any case, dreams were not vague musings and fantasies. They were real. In Barbara's family, the great American dream wasn't just left to the imagination, but lived out each and every day. Barbara's parents, William and Christine Mikulski, second-generation Polish-Americans, realized their dream in Willy's Market, across the street from the modest row house they called home. A few blocks away, Willy's mother and brothers operated Mikulski's Bakery, the first Polish bakery in Baltimore. It was there that Barbara's grandmother turned out her famous jelly doughnuts and her fragrant raisin bread.

The Mikulskis worked long hours in order to ensure that their three daughters had the benefits of a good education, and all of the opportunities it could provide them. The Mikulski girls were taught that they could do anything they dreamed of doing.

Barbara Mikulski grew up in this close-knit neighborhood, where power was never gauged in material goods but in the strength of the networks—relationships forged from the shared immigrant memory of the places left behind, and those newly built. Highlandtown was a "real neighborhood," Senator Mikulski recalls, where you could walk everywhere you had to go. It was also a community that lived and thrived on simple maxims. People helped people—not in an abstract way, but in daily practice. It was never described as charity. It was simply a given. If your neighbor had a need, you lent a hand. Willy and "Miss Chris" Mikulski regularly extended credit to their customers when American Can or Bethlehem Steel went on strike, or paychecks didn't stretch. Caring for others was as natural as breathing. It was a lesson that Barbara Mikulski took to heart.

Her role models were the nuns at the parochial school she attended. She found them to be well-educated women of remarkable inner strength, whose influence was reinforced and demonstrated every day. Mikulski adored the nuns, but she also had a rebellious streak, so she often found herself in trouble with them. Even so, she thought about becoming a nun, but quickly realized that she wasn't well suited for the role. "I wanted to be out in the world, to live my own life," she explains. Then, in an obvious understatement, she concludes, "I would have had trouble with the vow of obedience."

Nurtured by the support of her family, and inspired by the nuns at her school, the passion for service grew in Mikulski. The motto of the Christopher movement, a charitable organization for Catholic youth, became an essential part of her guiding philosophy: "It is better to light one small candle than to curse the darkness."

While political life—and certainly not the exalted position of a United States senator—was never specifically imagined, Mikulski was inspired by the social movements that came to dominate and define the sixties—in particular, the War on Poverty and the civil rights movement. She never thought of these movements as pie-in-the-sky exercises in idealism. They were as real as the ground beneath her feet. Grass-roots activism—taking care of one's own community—was the very essence of politics.

By the time Mikulski graduated from the University of Maryland School of Social Work, she had earned a master's degree in community organization. She already understood that community organization wasn't just the title of a course she'd taken or a subject that she'd majored in. It was an attitude that she exuded. It was the very ground of her being.

Mikulski's special brand of activism—a fiery devotion equally matched by a concrete tactical savvy—made her a tremendously effective social worker. Her mantra, which she repeats to this day, was: "You have to operationalize your good intentions."

By 1968, Mikulski was beginning to look beyond the neighborhood. "I decided I was going to go back to school and get my doctorate in public health so that when Ed Muskie became president, I could go to HEW and be made an assistant secretary or something," she recalls with a wry smile. Then fate intervened.

BARBARA MIKULSKI

I was all set to go when I got a call from a social-worker friend. She said, "You've got to come to a meeting tonight at the church." She told me about a plan for a sixteen-lane highway that would run through the neighborhood. She said, "They're afraid to fight the bosses, but your name is well known. Come and talk to them." So I went. And of course I stayed and fought. I figured it would take ninety days. It took two years.

I was outraged by the highway plan. The State Roads Commission called it progress, but how could it be progress? The highway would cut like a scar down the center of two neighborhoods. On the east were the Polish neighborhoods of Fells Point, Highlandtown, and Canton. On the west was the black community—the first neighborhood of black-owned homes in the city. These were vital communities. After World War II, the GI Bill had made it possible for the residents to buy their homes, to get educations. This was the new middle class, living the American dream. And it was going to be destroyed, just like that, by the State Roads Commission's new highway. Imagine the impact! Complete neighborhoods would be lost forever. And the residents would get nothing. No relocation benefits, and only the assessed value of their homes, which would have been about one-third of what they were worth. They were going to destroy Baltimore so there would be a highway for people living in the suburbs—that was really stupid.

At the meeting, people shared their fears and talked about their lack of power. The Democratic establishment, the mayor—everyone was behind the highway. What could they do? I pulled them together, told them, "We have to give ourselves a powerful name, so they'll take us seriously." So we created SCAR—the Southeast Council Against the Road.

Later, at a rally on the site where the highway would come through, I was asked to talk. I jumped up on a table, and I cried, "The British couldn't take Fells Point, the termites couldn't take Fells Point, and goddamn if we'll let the State Roads Commission take Fells Point!" Everyone cheered wildly.

But we had to do our homework. We had to build coalitions that would increase our power. The first step was joining forces with the black community on the west side. This was in the aftermath of the civil rights riots, and at that time it would have seemed impossible for the Polish neighborhoods on the east side and the black neighborhood on the west side to work together. The prevailing wisdom was that the European ethnics and the blacks were natural enemies. It certainly served the interests of the downtown powers-that-be to let that impression flourish. I knew differently. We were facing a common threat. I knew some of the leaders on the west side. They had formed a group called RAM—Relocation Action Movement. We had meetings, talked about working together. Naturally, there was some mistrust, but we were determined.

There was a hearing in West Baltimore one evening, and we left our neighborhood and took two busloads over. We drove to a school on Edmondson Avenue, in the black neighborhood. We walked inside. There were about sixty of us, and about four hundred African-Americans already seated. We signed up to testify.

The people of West Baltimore poured out their hearts about what this highway would mean to them—the destruction of their neighborhood, without any of the highway's benefits. Then a man from our group named Frank Milkowsky got up. Frank had worked on the docks, and he'd been a merchant mariner during World War II. He said, "My name is Frank Milkowsky, and I fought in World War II to save America. And I fought to save the government of the United States of America. Now I'm here to join with the black community, and the veterans and their wives, to save the neighborhoods of Baltimore."

There was thunderous applause. We clapped, we sang, we cheered. The ice was broken. We were together. Mutual need, mutual respect, mutual identification. That's how we created a citywide group called MAD—Movement Against Destruction. And the highway forces didn't know what hit them.

Today, Barbara Mikulski will be glad to show you the spot where the State Roads Commission abruptly stopped its highway project—just before it sliced in two the neighborhoods she fought to save. It remains a potent symbol of the collective power of an organized group that, to use Mikulski's favorite phrase, can "operationalize its good intentions." The highway fight led to Mikulski's first foray into politics. She decided that she'd rather be opening doors for others from the inside than knocking on doors from the outside.

Mikulski is often heralded as a mentor for women in government—a transformational leader whose goal is to teach others the skills to become leaders themselves. While she appreciates her position as a symbol of what women can achieve, she is not content to be merely a symbol. Her ideology is grounded in the practical. "When you go out and try to solve a problem, you have to have strategies to make it happen," she says. "If you only talk and don't produce, you are just one more disappointment. You contribute to the cynicism."

We are a nation of immigrants, shaped by the determination and sacrifices of those who came before us. Our immigrant legacy is the abiding symbol of our potential, as individuals and as a country. Like Barbara Mikulski, Senator Dianne Feinstein's incredible drive and commitment to creating a better world can be traced back to the influences of her immigrant heritage. Her mother's family was Russian Orthodox and fled St. Petersburg during the Russian Revolution, traveling by hay cart through Siberia and then by boat to Eureka, California. When Feinstein's grandfather died two months after their arrival, at the age of thirty-two, her grandmother was left nearly penniless, with four small children to raise and no job prospects. She didn't even speak English.

Feinstein's paternal grandfather was born to a family of Polish Jews, living in a small town that bordered Russia, where they were subject to terrible pogroms. He fled at the age of fourteen to avoid conscription into the Polish army, which would have brought its own special terror, and stowed away on a ship headed for Boston. When he was nineteen he made his way to northern California, where he met and married a young Lithuanian woman. They settled in San Francisco, where they had five children prior to losing their home and business in the 1906 earthquake. They moved across the Bay and had six more children. Because the family was poor, only half the children were able to go to school, while the others worked. Feinstein's father was one of the lucky ones. He was designated to be educated. He became a physician, and eventually would be appointed the first Jewish full professor at the University of California Medical Center.

DIANNE FEINSTEIN

Imagine my family's pride at that accomplishment. My uncle Morrie, who helped put my father through school, had very little formal education, but he, too, was part of the American dream. He became a successful businessman with an abiding interest in politics. It was my uncle who really gave me my desire to participate in society. On Monday afternoons, he would take me with him when he went down to the board of supervisors—only he called it "the board of *stupidvisors*." He would always say to me, "Dianne, you get an education, and you do the job right."

The message of the importance of education and the opportunities it created was reinforced repeatedly, which is one reason I have always made education such a priority. It really is the American dream. I realized that it was a direct legacy of the sacrifices my grandparents and my parents had made. But I had another legacy as well. I was born into an American family during the Holocaust, so while I learned about the possibilities that existed as part of the American dream, I also grew up understanding that terrible injustices could be inflicted on people because of hatred and bias. That awareness made a strong impression on me and instilled in me a commitment to the social good. It was my obligation to stand up against injustice and to fight for the rights of every person—regardless of their race, creed, color, sex, or sexual orientation—to live a safe, good life.

For a time I thought I might follow my father into medicine. I adored him, and he wanted me to be a doctor. It had meant so much to him and to his family when he received his medical degree. But in my first semester at Stanford, I got a D in genetics. I came home and said, "Dad, I don't have the aptitude for this. I can't possibly be a doctor." And then he told me the first and last lie he would ever utter: "Don't worry, I got a D, too." He was trying to encourage me not to give up, but I knew that my father had never seen the south side of an A.

I realized I would never be a doctor, but then I discovered something I loved and which I was good at. I took a course in American political thought, and I wrote my heart out in that course. It was the beginning of what would be my life's work.

After college, I became a Coro Foundation intern in San Francisco, which gave me an opportunity to do a team project in the postconviction phases of the administration of justice. I sent the paper I wrote to then-Governor Pat Brown, because I knew it was good, and I believed he could do something with my ideas. He did. He appointed me to the California Women's Board of Terms and Parole. I was the youngest member of a parole board in the United States, and it was a heavy obligation. But I compensated for my youth by doing my homework. I worked hard, and I took my job very seriously. I also served on the Crime Commission, which was the jail investigative body. I wrote newspaper articles and gave speeches. And people responded to me. So I thought it was time I did what my uncle Morrie had always wanted me to do. I ran for election to the board of supervisors.

Not only did I win, but I topped the ticket, which entitled me to be the board's first woman president. Some people had a hard time with that. There were editorials in the newspapers saying, "We can't have a woman as president of the board of supervisors—especially a woman who has never served in an elected office. She should do the statesmanlike thing, and yield to the next highest vote getter."

So I looked around to see who that was. It was a man, of course—a real estate broker. I decided to keep the position, and I made a success of it.

I loved the work. I have never served in an office that I didn't love. I always tell young people to find the thing that they're good at and which they love. If they're lucky, they'll want to do it for forty or fifty or sixty years. And there are bound to be defeats. But if you're doing what you love, you can handle the defeats, and get up the next day and come back, still loving it.

What does it mean to pass the torch to a new generation of women leaders? Each of the women senators understands that at any given moment, she could have a substantial impact on someone's life. She doesn't always know when such moments will occur, but occasionally she is later informed.

"It's a wonderful feeling," Susan Collins says, "to learn that you've made a real difference in a young person's life." As an example, Collins mentions a young woman on her Washington staff named Emily Woodman. Collins first met Emily in 1994, when she spoke to Emily's class at Bangor High School during her campaign for governor of Maine. After hearing Collins speak about the difference one person can make, Emily volunteered to work on her gubernatorial campaign. When Collins lost the general election and moved to Bangor, Emily would sometimes see her around town. In 1996, when Collins announced her candidacy for the Senate, Emily was among the first to volunteer to work on her primary campaign. She became a staff member during the general election, which Collins won. Emily's experiences on Collins's campaigns inspired her to choose political science as her major in college. During her senior year at the University of Southern Maine, Emily interned in Collins's Portland office, and when she graduated in May 1999, Collins hired her to work in her Washington office. It was the first time Emily had ever been out of Maine.

When Emily arrived in Washington, she told Collins, "I'm here because of you. You changed my life."

"No matter what else I do in my life, having an effect on young women like Emily is a legacy that makes my hard work worthwhile," Collins says.

SUSAN COLLINS

In a sense, that torch was passed to me by our former senator from Maine, Margaret Chase Smith. Although I did not realize it at the time, a conversation I had with Senator Smith when I was eighteen was the first step on a journey that would lead me to run for her seat in the United States Senate twenty-five years later.

When I was a senior at Caribou High School, I was one of two students selected statewide to participate in the United States Senate Youth Program, sponsored by the William Randolph Hearst Foundation. The program included a scholarship and an exciting weeklong program in Washington, D.C. For a high school student from Caribou, Maine, who had never been on an airplane before, this was an incredible opportunity. Student delegates from all over the country listened to cabinet officials and members of Congress, toured monuments and museums, and met briefly with their senators—if the students were lucky and if their senators had time. I arrived at Senator Smith's office a bit breathless about meeting her. The senator's staff ushered me into her private office, where Senator Smith, wearing her trademark red rose, stood to greet me. Much to my surprise and delight, she then spent nearly two hours talking with me and answering my many questions. She told me about her service on the Senate Armed Services Committee, the importance of a strong national defense, and the definition of "full employment," which had been the subject of much debate at the time. But what I most remember was her urging me always to stand tall for what I believed in. She described her own famous "Declaration of Conscience," in which she spoke out on the Senate floor against McCarthyism. As I left her office with a copy of her speech, I remember being so proud that Margaret Chase Smith was my senator. I also recall thinking that if she could be a United States Senator, women could truly do anything!

3

Don't Get Mad, Get Elected

I'm Nobody! Who are you?
Are you Nobody—too?

EMILY DICKINSON

Senator Patty Murray will never forget the moment when a member of the Washington State Legislature let her know, in no uncertain terms, that she was nobody, that the issue she cared about didn't matter. She had no clout.

Patty Murray was a young wife and the mother of two preschoolers in 1980, living in Shoreline, Washington. With a bachelor's degree in recreation from Washington State University, Murray set her goal to teach young kids. All of her carefully laid plans changed in an instant the day she was casually dismissed by that legislator. The course of her life was altered forever by a boorish remark.

PATTY MURRAY

Politics was the farthest thing from my mind. I had never been to our state capitol. I was working as a parent volunteer for the Shoreline Community Cooperative School, which was a wonderful parent-child education program sponsored by the local community college. My kids attended the preschool program, and I took courses and did volunteer work. One day I went to class and the teacher announced, "I'm sorry. The program is going to end. The state legislature is taking the funding away." I was shocked. I said, "Wait a minute. Who are they to decide this? This is a fabulous program." I assumed that the legislators were just uninformed. If they knew how valuable the program was, they would never cut the funds. Oh, the beauty of naïveté! I look back now, and think, "Oh, God, what ever possessed me?"

I decided to go to the state capitol in Olympia, to talk to some legislators and convince them of their mistake. I didn't even think twice about it. I put my two kids in the car, and off we went. At the capitol, kids in tow, I started to talk to everyone I could find—legislators, staff, lobbyists. Everyone was very polite, but I didn't feel as if I was getting anywhere.

Finally, one legislator I'd pigeonholed listened to my story, and then let me know what he really thought. He crystallized my problem. "Lady, that's a really nice story, but you can't get the funding restored," he said. "You can't make a difference. You're just a mom in tennis shoes." And I could see that was how he honestly viewed me. I was not a policymaker, I was not a lobbyist, I was not somebody who knew what I was doing. I couldn't effect change. I was just a mom in tennis shoes.

I was speechless. I was also furious. I couldn't make a difference? I was just a mom in tennis shoes? What's wrong with being a mom in tennis shoes, anyway? I was a voter. I lived with the policies. I understood them. I knew this program. But according to a bunch of guys in suits over at the capitol in Olympia, what I cared about wasn't important. I seethed all seventy miles back to Seattle, and by the time I arrived home, I had made a decision. I wasn't going to let them get away with this.

So I did what every mom does. I got on the phone. I called all of my friends. I called every instructor in my community-college program, and all of the other instructors across the state. I got the names of all the parents in every program, from Yakima to Spokane to Aberdeen to Bellingham to Seattle. I started making calls. I said, "Hello. My name is Patty Murray. You've never heard of me, but here's what the state legislature is doing to the program that you and your kids are in. Will you help me? If you can't help me, could you give me the names of three more people?"

I literally called hundreds of people, and put together a list of about fifteen thousand parents. These were moms like me, and they responded like I did. "You bet I'll help." We wrote letters and made phone calls, and we organized a big rally in Olympia, on the steps of the capitol building. We showed up at every hearing. We were a presence. And we were heard. It took almost a year, but we got the program reinstated.

That's what got me involved in politics. I understood for the first time that the decisions government made had an impact on me. And I figured that I could sit at home and say, "Oh, well, that's too bad," or I could get involved and be a part of the decision-making process.

———————————————————————

By the time that Patty Murray decided to make a run for the United States Senate in 1992, she was no longer being dismissed as "just a mom in tennis shoes." Still, the label became something of a populist rallying cry. Each of the women who made their way to the Senate had traveled a long road to credibility—a road paved with involvement with hundreds, if not thousands, of grass-roots groups—from neighborhood watches, community child-care centers, citizen action groups, and on to local school boards, city councils, county legislatures, state offices, and the Congress. The women of the Senate weren't anointed by the high priests of their local political parties to the office they attained. They paid their dues. As Murray puts it, "We've tried to rebuild our communities, and now we're trying to do the same with our country."

Each woman's path to power has been a long and circuitous one. A lot of the time, it doesn't remotely resemble anything that smacks of politics. One can't imagine a male politician organizing a bake sale or a block party. But the training grounds of the block, the school, the church, the synagogue have proved to be amazingly effective. "How do you pass a bill?" Murray asks. "You talk to people. You listen to them. You reach out. You work hard for what you believe in. It's simple, really."

Simple, maybe. Easy, never.

In their fight for credibility, women have an especially hard time shaking off labels that attempt to define them in dated, stereotypical ways. Although Kay Bailey Hutchison had been a state representative, state treasurer, banker, and small businesswoman when she was elected to the United States Senate in 1993, she was greeted after the election with a newspaper headline: FORMER UNIVERSITY OF TEXAS LONGHORN CHEERLEADER ELECTED . . .

Former senator Carol Moseley-Braun also received the cheerleader treatment when she was elected to the Senate from Illinois in 1992. Moseley-Braun was a lawyer, a former federal prosecutor, and a veteran state senator before taking her Senate seat. This was a formidable string of accomplishments by any measure, but the senator couldn't escape comment on her more feminine attributes. *The New York Times* described her as "a den mother with a cheerleader's smile."

Barbara Boxer still remembers campaigning for county supervisor in 1972 and giving an impassioned speech about the environment. When she asked for questions, a woman in the audience waved her hand excitedly. Boxer was thrilled to have elicited such an enthusiastic response until the woman said, "Tell me, Barbara, when do you have time to do your dishes?"

Women's fight for credibility is complicated by the selective amnesia of the media, which can all but ignore a woman's record in the interest of feminizing her. In 1992, when Dianne Feinstein was running for the Senate, a reporter asked her if she thought a woman could be as tough as a man when it came to making hard public policy decisions. She locked her eyes on his and replied, "Do you believe that toughness comes in a pin-striped suit?"

Surely any questions about Feinstein's toughness should have been answered long before 1992, during her impressive career in government, which included serving on the parole board, on the Crime Commission, eight years on the San Francisco board of supervisors, and two terms as mayor of San Francisco. Her definition of toughness is tenacity: "You have to earn your spurs."

DIANNE FEINSTEIN

If I were advising young women today about becoming involved in public life, I'd tell them, "Start on the school board, go for a spot on the town council. *Earn your spurs.* That's the key. Don't flit around like a moth—light and leave, light and leave. Develop a portfolio of expertise—something you're really good at, so that people will turn to you. Develop your credibility, your integrity, show people they can trust you. Make a contribution. Be effective, because once you show you are effective, you're a force to be reckoned with."

My portfolio of expertise was criminal justice. I sat on the parole board. I was on the Crime Commission. I could write, and my writings would be published. I was asked to speak. People sought my advice. I had something to add to the debate that was of consequence.

This is exactly the path that Patty Murray took. Murray built her political credibility from the ground up, beginning with the school board, where she served for four years, including a year as its president. Still, by the time she ran for the state senate in 1988, even Murray's supporters feared that she was perceived as not having the necessary weight, the gravitas, to win.

PATTY MURRAY

I was told that I hadn't been in line long enough, that I didn't have the history, that I couldn't raise the money. Believe it or not, I was told many times that I was too short. I was actually advised to run as *Pat* Murray, so that people might think I was a man. And I said, "Wait a minute—I'm running *because* I'm a woman. I'm running because we *need* women." And they didn't want people to know I had young kids because their reaction, of course, would be "Oh, my God, how are you going to take care of your kids?" I said, "I'm running *because* I have young kids. We need policymakers who understand what women are going through so these policies work for women." I was angry. Imagine! They wanted me to trick voters into thinking I was a man. I said, "No way. If I get this job it's because I am a woman and I want people to know I'm a woman."

Defining Moments—Washington, D.C., 1991

*Calling 1992 the Year of the Woman makes it sound like
the Year of the Caribou or the Year of the Asparagus.
We're not a fad, a fancy, or a year.*

BARBARA MIKULSKI

The courage to stand your ground over the important issues and negotiate others is a tricky matter. Once you are defined by a moment's action, you can carry it with you forever. For Barbara Boxer, that defining moment came in October 1991, when she was serving in the House of Representatives. Her decision to take a highly public stand on an issue that pitted the perceptions of male legislators and those of female legislators against one another solidified her reputation as a feminist activist—a tag that can be trouble for women in elections. The issue was the confirmation of Clarence Thomas to the Supreme Court.

Anita Hill, a distinguished law professor who had worked at the Equal Employment Opportunity Commission while Judge Thomas ran it, came forward and told the Senate Judiciary Committee that Thomas had sexually harassed her while she worked for him. Boxer was relieved by Hill's candor. She already had serious reservations about Thomas's ideological positions on abortion and economic justice issues, and she hoped that in light of the charges, President Bush would withdraw Thomas's name and present a new, less conservative African-American nominee. When Bush announced that he was sticking by his candidate, and the Senate Judiciary Committee stated it would not air the sexual harassment charges, Boxer was stunned. So were many of her female colleagues in the House, who created a stir by bucking the convention that members of the House did not comment on matters that were before the Senate. The incident gave rise to a dramatic scene that was memorialized on the front pages of newspapers across the country: seven Democratic women from the House of Representatives walked over together from the House to the steps of the Senate, demanding to be heard.

BARBARA BOXER

Our actions have always been portrayed as those taken by a group of renegade women, but that's not what it was about at all. That was an angle tailored by the media. We wanted to try to help the senators rethink their position, and we felt that they might appreciate hearing our perspective. Remember, at that time there were still only two women in the Senate—Nancy Kassebaum and Barbara Mikulski. And neither of them was on the Senate Judiciary Committee. All we were asking was that the Senate take a serious look at Anita Hill's charges. We assumed that our Democratic colleagues would be more than happy to hear our views. They were discussing what to do at that very moment, at their regular Tuesday Democratic caucus lunch. We went up to the doors of the room and politely asked if we could have a few minutes to come in and share our views. The senior staff members guarding the doors wouldn't let us in. Did they realize that we were seven congresswomen? Yes, they knew. Did they realize we were colleagues from the Democratic side of the aisle? Yes, they knew that, too. We were turned away, failed supplicants.

It was humiliating to be so summarily dismissed—to have to beg for a hearing. But we kept demanding to be heard until Senate Majority Leader George Mitchell agreed to meet with us. And so we got our hearing. It turned out to be a travesty. It was shameful. But it was also a wake-up call. All over the country, women watched those hearings, and reacted to seeing that long row of white male senators. It was made painfully clear that they just "didn't get it."

I was already running for the Senate when the Clarence Thomas hearings took place. I have to say that many people—even my supporters—asked me why I had to be so outspoken during that time. They thought it would hurt my chances for election. I don't believe that anyone was prepared for the huge backlash the Thomas hearings caused—the unleashing of the women's voice in that next election. Women saw what happened, and realized they still weren't receiving anything approaching a full measure of representation. They were saying en masse: "Who is going to represent my interests? Who will speak for me?" And so they voted to ensure that they might at least have a chance to be heard.

Patty Murray hadn't considered running for the Senate in 1992 until the Clarence Thomas hearings. She recalls sitting in her living room, watching the hearings and looking at the row of men asking the questions. "I kept thinking, 'Who's saying what

I would say if I was there?' No one was expressing my opinion. I didn't feel as if my voice was being heard. Later, we went to a party in our neighborhood, and all of the women were talking about the hearing, and they felt the same way I did. And right in the middle of the party, I said, 'You know what? I'm going to run for the Senate.' They looked at me and said, 'Uh-huh.' Like, 'There she goes again.' But I told them, 'If nothing else, a year from now you'll be able to vote for somebody that you believe will represent your position.'"

The 1992 election came to be referred to as the Year of the Woman. It brought four additional women into the United States Senate—Barbara Boxer, Dianne Feinstein, Patty Murray, and Carol Moseley-Braun. But the women elected to office that year weren't entirely comfortable with the "Year of the Woman" label. It seemed to reinforce the stereotypical notion that they had been swept into office on a whim: a popular, if temporary, surge of enthusiasm for women by women. It underscored the perception that women who achieved public office were there to fulfill short-term goals, rather than being around for the long haul.

"In many ways, the Year of the Woman rhetoric was demoralizing," Barbara Boxer confesses. "People kept saying, 'You won your Senate seat because it was the Year of the Woman. Wait until you run for reelection. It won't be the Year of the Woman then, and you'll lose.' Fortunately, that didn't happen in my case, but the sentiment ran strong. I had to fight a tough race for reelection. Besides, there was something pathetic about it. We increased our numbers from two to six women in the United States Senate, and *that's* what everyone called the Year of the Woman!"

Republican women agree that these labels diminish women's contribution, no matter which side of the aisle they're on. Kay Bailey Hutchison won a special election in 1993 to fill the seat vacated by Lloyd Bentsen when he became President Clinton's secretary of the treasury. She then won reelection in 1994, as did Nancy Kassebaum. Olympia Snowe was also elected to the Senate that year. When the press tried to label 1994 as "The Year of the *Republican* Women," the newly elected senators rejected it outright. They felt that it was an attempt to diminish their achievement. "Don't forget," one senator said wryly, "that 1994 was also called 'The Year of the Angry White Male.' Which of our colleagues wish to claim *that* label?"

Perhaps the greatest hurdle to overcome in achieving public credibility is finding the opportunity to prove—*to yourself*—that you can be effective. Confidence isn't enough. Believing you can do something, and actually doing it, are two very different things. It's necessary to go out there and do it, to have an effect, make an impact. That's the true measure. When Kay Bailey was first elected to the Texas State Legislature, she knew she had a lot of proving to do.

KAY BAILEY HUTCHISON

When I first came into the legislature, I was young and single, and I had to prove that I could handle the job—that I could represent the people from my district, and do it well.

There was always the feeling, particularly among men, that because I was a woman, maybe I wouldn't be tough enough. "She's going to go soft," they'd grumble. "She *says* she's for strict budget controls, and she's for the death penalty, and she's against raising taxes. Well, we'll see if she's tough enough to stick to it." Overcoming that kind of undercurrent was one of my greatest challenges.

Suspicions were also aroused because I was a Republican. At that time, women interested in politics were associated only with the Democratic Party. There were four other women in the Texas House, and they were all Democrats. All of my friends from college were also Democrats, and they were just baffled. "How did you become a Republican? Don't you know that in Texas, if you want to have a chance to win, you have to be a Democrat?" Well, I just had a gut feeling that the Republicans had the better vision for the long term.

When I joined the Texas legislature, it was just at the beginning of the Republican Party in Texas, which had always been a staunchly Democratic state. It was also at the beginning of the women's movement. I was the only conservative woman in the House, so sometimes the four Democratic women would vote one way, and I'd be on the other side, voting the other way. But one of the things I loved most about being in the House was that so often our work was about real issues that transcended all of the partisan differences. That was especially true when it came to fighting for issues that grew out of women's lives. My first experience of women bonding for a purpose was when the five women in the House sought to give rape victims the same protections as other victims of crimes. Because the Democratic women worked their side of the aisle and I worked mine, we passed a law that became a model for other states to follow.

We addressed the basic inequities that were inherent in the current system. The way that rape victims were treated had to change. More often than not, rape victims who had the courage to file charges and follow through on them were ultimately forced to relive the details of their rapes in front of an entire courtroom. They had to take the stand during these trials, and then had to defend their actions! They were allowed to be extensively cross-examined by defense lawyers. Victims were being mercilessly degraded and traumatized. They were the ones being put on trial, and it wasn't right. Our bill stipulated that a woman's background could not be brought out in a rape trial, except *in camera*—that is, outside of a jury's presence. If a defense lawyer wanted to make the case that a woman's background was relevant, the judge would consider it in closed session, so as not to taint the jury. It would not be admitted into open court unless the judge deemed it relevant. A lawyer couldn't just say, "When this girl was in high school, she had a reputation for being promiscuous," or, "She was flirting with the men in that bar." We also increased the statute of limitations to put rape in line with other crimes.

I noticed that when the women of the legislature brought issues like this to the other legislators' attention, the men were almost always resistant, at least initially. Why? Frankly, they'd just never thought about it. But when we had a chance to talk about it with them and put it on a more personal level, they came around. When we pointed out that it could be their sister or their daughter or their wife, they suddenly got it. It was like lightbulbs going on over their heads. So when I'm asked, "Does it make a difference that women are part of the process?" I say, "You bet. We bring our life experiences to the table." Nobody fought for homemakers to have retirement accounts until we did in the Senate in 1993, for God's sake. And here were all of these older women, living longer but earning smaller pensions, trying to get by with less security and fewer benefits. When a woman has worked inside the home, what is she to do when her husband dies or leaves her after a thirty-year marriage?

It wasn't that men were against these changes. They just hadn't considered the issue before because they hadn't experienced the problem in their own lives. As women have become a part of the system, that's changing.

In 1990, when Dianne Feinstein and Pete Wilson were engaged in a hotly contested race for governor of California, they faced off in a televised debate. The next day, the *Los Angeles Times* declared the debate to be a virtual tie. However, the paper gave Wilson a slight edge because he "looked" more like a governor.

What does a governor look like? What does a United States senator look like? The Constitution has a simple answer: government representatives look like *you—the people.* For most of our history, that has meant white males. But today, eighty years after women won the right to vote, we are gradually coming to terms with the idea that a representative government must open itself to the diversity of its citizenry. Few would disagree with this ideal, but the reality is more difficult for many people to support.

We are, after all, the products of our histories, and the belief systems and biases that have been coded into our operational hard drives. The movement from likeness to diversity requires a paradigm shift as fundamental as the one that spurred the birth of our nation. For to accept diversity, we must also accept the change it brings to our entire notion of how to govern effectively.

Barbara Mikulski once observed that the history of the United States is often framed around its wars and battles rather than its social movements. Yet, she noted, it has been the social movements—ordinary people coming together to address felt needs or correct social injustices—that have been the heart and soul of the revolution that is America. Wars and battles occur intermittently to shock the status quo. Social movements represent the long haul—the slow, steady wearing away of negative practices, and the gradual rebuilding of a positive foundation. Both are necessary. However, whenever we look at a major social movement, we see women in the front ranks—a force of stability and constancy.

Women do not have to be like men to succeed. Rather, they need to appreciate and respect the importance of their contribution. Sometimes that is simply a matter of pointing out to them that the work they're doing gives them the skill to do more.

Mary Landrieu grew up in a large and lively political family. Her father was the widely respected but controversial civil rights leader and New Orleans mayor Moon Landrieu. By the time she reached adulthood, she had a tremendous amount of self-confidence, fostered not only by her father but also by a strong mother, who was a leader in her home and in the city. But Landrieu never considered going into politics. In her world, women were "unofficial" leaders. When she was encouraged by friends to run for office, she learned that she could win. But she also learned that winning wasn't enough when it came to establishing credibility.

MARY LANDRIEU

I was volunteering on a campaign for a friend of the family. Some of my friends and fellow campaign workers came up to me and said, "You're very good at this. Have you ever thought of running yourself?" Of course I hadn't. But suddenly I found myself thinking, "They're right. Why not?" What gave me the confidence was that I had already been elected by my peers to student body government offices and my district was like a big neighborhood—only about fifteen blocks wide and forty blocks long with about twenty thousand voters. So I started going door to door. I was only twenty-three years old, so I fully expected that I'd have to give a vigorous defense of my age. Instead, my neighbors were pleased to see a young person involved in politics. They'd say, "That's great. I'd given up on the younger generation." I won the election because I campaigned on issues people cared about. One of these was drainage. In May 1978, our district had suffered terrible flooding. Inadequate drainage led to many people's houses being underwater. I promised to do something about it. After many years I was able to get the drainage system redesigned.

I would be understating it to say I was *shocked* to come face-to-face with the reality of how few women there were in the

Louisiana legislature when I walked through the door in 1979. I'd never really thought about it until I got there. I'd grown up in a home where boys and girls alike were raised with high expectations. We were encouraged to excel academically and to take on leadership roles. We all played competitive sports. Our parents set an example by being equal partners in our home, although their roles were different. In addition, my high school, Ursuline Academy, was all girls, so there were no limits placed on what we could do. That's why I was taken aback when I walked into the legislature, and all I could see were male faces. Not just the other representatives, but their staffs and most of the press were male. It didn't seem as if they were expecting me or anyone who looked like me.

This was the first time in my life that I was in a place I wasn't sure I should be. It was challenging enough to learn the job without having this added pressure. I experienced initial tension and frustration because my issues were so different from those of the body at large. Today, the political establishment is interested in issues related to education, child care, child abuse, domestic violence, women's rights, and the rights of the disabled. These issues weren't on the front burner in Louisiana in 1979.

My eight years in the legislature were tough but very valuable. I'd had such high expectations of what I could accomplish, but after two terms, I felt I had accomplished what I could. I decided to make a change. The job of state treasurer happened to be held by a remarkable woman named Mary Evelyn Parker. She had been a pioneer—getting elected to a state job in a male-dominated field. She'd done a good job in office, and people had become used to the idea that it was a woman's job. When she announced that she was not running for reelection, I realized that it would be a great opportunity to serve the state from a different platform. Using what I had learned in the legislature and building on my experience as a member of the Appropriations Committee, I entered the race. I beat a formidable field of men, winning when few expected I could.

In the state treasurer's job, I found my niche. I built a reputation for being an independent thinker—someone who was an outspoken and effective advocate for fiscal responsibility and responsible government. Finally, I knew I was making a difference in helping our state move forward.

Critical components to women achieving credibility and being elected to office is party support and money. You get party support by showing that you can win. You show that you can win by demonstrating that there is public enthusiasm for your candidacy. You demonstrate public enthusiasm for your candidacy by getting your message out early in the game. And that takes money. Since 1985, many Democratic women's candidacies have been energized in the beginning, when it mattered most, with the help of a remarkable women's fund-raising group called EMILY's List—which stands for "Early Money Is Like Yeast." Started in 1985 by Ellen Malcolm, EMILY's List has been responsible for making Democratic women more competitive at all levels of government. More recently, Republican women have started developing independent fund-raising resources as well. The WISH List (Women In the Senate and the House), for example, backs pro-choice Republican women for office and proudly touts its record of backing each of the Republican women currently in the Senate. Likewise, organizations such as the National Federation of Republican Women have played key roles in advancing leadership by Republican women.

While polls show that confidence in women candidates is on the rise, their numbers have not increased substantially during the past five years. In 1999, fourteen women's groups united to found the Women's Information and Resource Candidate Clearinghouse, a bipartisan organization whose goal is to bring more women into politics.

"We'll have equity when gender is no longer an issue," says Mary Landrieu. "When we're really measured on the basis of brains and talent, we'll do just fine."

4

Triumph and Disaster

*Out of my experiences in life has grown a conviction that
no pursuit is as valuable as, or worthier than, the simple idea of
helping others—of enabling individuals to improve their lives, to
soften the hardest days and brighten the darkest.*

OLYMPIA SNOWE

To choose a life of public service is to give up the comfortable il-
lusion that all is right with the world. It is to take responsibility
for trying to make things better, even while knowing that you'll
never be able to do enough. It is the constant process of mold-
ing chaos into purpose. It is remaining optimistic while knowing
that life holds equal shares of triumph and disaster. And sometimes, it is being called upon to summon inner resources you
didn't know you had in order to hold together the collapsing
spirit of your people.

November 1978—San Francisco

Dianne Feinstein's future was changed forever by an act of insanity. If there is a moment of truth in each life that illuminates one's destiny, this was surely hers.

On the day that disgruntled former San Francisco city supervisor Dan White assassinated San Francisco mayor George Moscone and supervisor Harvey Milk, Dianne Feinstein, president of the city's board of supervisors, became the acting mayor of San Francisco. Her strength and steadiness calmed the jittery nerves of the city in the hours and days following the murders. Her behavior cemented her reputation and thrust her into the national spotlight.

Ironically, the crisis occurred at a time when Feinstein's inner resources were as depleted as they had ever been. She was grieving over the death of her husband, Bertram Feinstein, from cancer. She was demoralized by two unsuccessful runs for the mayor's office, and she wasn't sure that she had a future in politics. Exhausted and disheartened, Feinstein had started informally letting friends and supporters know that she wouldn't seek reelection when her term was up. She was considering getting out of politics for good. Then tragedy intervened.

Feinstein was in her office at city hall that fateful day, when she saw Dan White run into Milk's office and shut the door. When she heard the shots, her first thought was that White had committed suicide. She immediately realized there had been too many shots for that. The following moments are burned into Feinstein's memory, like a nightmare in slow motion. She turned the knob of the closed door and walked into Milk's office. The city supervisor was lying on his stomach. She knelt down beside him and tried to find a pulse, but instead, her finger slid into a bullet hole in his wrist. He was dead. Within minutes, Feinstein was informed that the mayor had also been assassinated. She was now in charge. She was the acting mayor.

When Feinstein appeared before the television cameras and recounted the horrific details of that day to the people of San Francisco, her voice often shook with emotion, but her steadiness of purpose never wavered. Drawing from a deep reserve of strength and compassionate resolve, she was able to explain what had occurred and calm all fears. She communicated that, though sorely diminished by the deaths of George Moscone and Harvey Milk, San Franciscans would rally together and overcome this tremendous blow to the heart of their beloved city.

Even Feinstein's critics have always acknowledged that her strength and resolve in the days following the murders helped the city to remain calm and eventually heal. Thrust into a position where she might well have given herself over to despair, Feinstein instead set aside her own doubts and fears in order to tend to the needs of the people of San Francisco. "Forged in tragedy . . ." the opening line of a powerful campaign ad that later drew on that experience, became a defining aspect of Feinstein's political life.

This ability to focus on the greater need, the public good, is the defining characteristic of a dedicated public servant. The mantle of public service can be a heavy one, and often requires a sublimation of personal feelings and a certain stoic reserve. A woman who dons the mantle is placed in an especially difficult position. Women in public life are never forgiven for their tears, but they are held in suspicion for displays of stoicism. Women are still forced to ride the ever-shifting teeterboard of proper feminine behavior. They must fight notions that they are not tough enough while also taking care that they don't appear to be too tough—that is, too much like a man. Throughout her political career, Feinstein has been criticized for both, but she has learned not to waver in the face of criticism. She knows herself, and she knows what she can do. She has been tested by trial more than most public officials, male or female, and she has taken its lessons to heart.

DIANNE FEINSTEIN

It was as if the world had gone mad. Our city had just started to deal with the horror of the Jonestown mass suicide, which involved nine hundred San Franciscans drinking cyanide-laced Kool-Aid, when we were hit with this enormous blow. As the new mayor, I, too, was deeply shocked by this. I was also rather staggered by the fact that I was now replacing the man who had defeated me in the last mayoral election. I made a decision that the goals of the assassin would not succeed completely. I kept all of the mayor's staff, and I followed through on all his initiatives for the remaining year of his term.

I also had to decide right away whether or not I would run for reelection. I realized that I could not let myself become a lame duck. Like it or not, I had no choice. The city needed to be reassured that there would be some consistency as we put the broken pieces back together.

The process of doing that was a fully nonpartisan effort. Our commitment was to the city, beyond all party considerations.

From that nonpartisan experience, I drew my greatest political lesson—the heart of political change is at the center of the political spectrum. The more diverse we become as a society, the more effective government can be when it operates from the center. It's not about being "politically correct," but about what works and solves the problem at hand. If you work from the center of the political spectrum, you can listen to the right, you can listen to the left, you can make judgments as to what is the best thing for all the people. And I did that for nine solid years as mayor of San Francisco.

There were more crises looming. In 1981, I began to hear rumors about a "gay cancer," and I instructed the director of public health to investigate immediately. He confirmed that the Centers for Disease Control in Atlanta was reporting the outbreak of a new and deadly disease, which afflicted gay men. Little did any of us know that we were in the early days of a plague that would create havoc in our city and across the nation. In the coming years, I would work closely with leaders in the gay and lesbian communities to fight back on every possible front. We spent more money for AIDS than any other city. We opened the first AIDS wards, conducted research, facilitated clinical drug trials, and adopted tough antidiscrimination measures. I was proud of our work, and I believed that we would be able to end the epidemic within a few years. That didn't happen, and it was my most profound disappointment. I, too, watched many dear friends and colleagues waste away and die during those years, and it is a terrible tragedy that we have not found a cure twenty years later. I have continued my commitment to lobby for research funding and to keep the issue on the front burner.

If you're serious about being part of government, you have to be prepared for the long haul. You have to be prepared to lose many of your battles the first, second, and third time around, and then get up and go back to work for the fourth time. Sometimes you'll get it on the fourth time around. Sometimes it'll take fifty tries. Some things will still be unfinished when you die. But if you keep at it and lay the foundation and bring others in, they'll carry on.

As Feinstein discovered, every trauma presents a clear choice: are you going to be the one to fold or the one to stand tall? It takes tremendous courage to make the latter choice, but the willingness to do so distinguishes one as a leader. Senator Olympia Snowe of Maine knows about this. Her life has been punctuated by a series of tragedies that have tested her mettle and left her in a solitary spotlight, requiring her to draw upon the strength of her own inner resources.

Augusta, Maine

The state of Maine is known for its rugged, rocky coasts, its pristine forests, lakes, and mountains, and its breathtaking New England beauty. Its people are thought to embody the very essence of the sharp Yankee—a soft-spoken, even stoic lot. Former Maine governor, U.S. senator, and secretary of state Edmund Muskie was fond of saying that he spoke only when he felt he could improve on the silence. But it is more than stoicism that is reflected in the hearts of the people of Maine. There is a communal spine, a strength that comes from living in the northernmost state on the East Coast. Thriving through Maine's crisp autumns and harsh winters, its relatively short springs and summers, serves to underscore the population's unflagging tenacity in the face of challenges that might daunt others. It was a fitting place for Olympia Snowe's parents, George and Georgia Bouchles, to live and raise their son and daughter.

George Bouchles was an immigrant from Greece, and Georgia Goranites was a first-generation Greek-American girl. They married and settled in the small city of Augusta, Maine's capital. George and Georgia operated the State Street Diner, just down the street from the state legislature. Georgia loved politics, and the diner was frequented by legislators and others with business at the state capitol. "I still encounter people today who remember me as a preschooler from the restaurant or hanging around the secretary of state's office, where my mother and I used to visit her friends who worked there," Snowe says.

Later the family relocated to nearby Lewiston, a mill city with a small but vibrant Greek-American community. Olympia's early years with her parents and her brother, John, were rich with love and grounded in the faith she learned from the Holy Trinity Greek Orthodox Church in Lewiston. Olympia was taught early by her parents the value of hard work and to put the needs of others before her own.

This spartan code would stand her in good stead when the first tragedy of her young life occurred. In July 1955, eight-year-old Olympia lost her mother to breast cancer. Her father, in poor health, struggled to raise the two small children alone.

In April 1956, when Olympia was in the third grade, her father realized he was unable to care for the children himself and sent her to St. Basil's Academy, a school for girls run by the Greek Orthodox Church in Garrison, New York.

That November George Bouchles died of heart disease. Nine-year-old Olympia and her thirteen-year-old brother, John, were orphans. Arrangements were made for John to live with relatives and Olympia to continue at St. Basil's during the school year. When she was home in Maine, she was cared for by her uncle Jim and aunt Mary Goranites, the brother and sister-in-law of her mother. Uncle Jim was a barber, Aunt Mary worked in a textile mill, and they already had five children of their own. Like Olympia's parents, Aunt Mary and Uncle Jim believed in the value of hard work and a good education.

Today Olympia Snowe's external serenity and easy smile belie the tragedies that have shaped her life. She has taken the fundamental values of family, faith, work ethic, and education and added to this foundation the further measure of her own caring and compassionate nature. "From the time I was young, I always felt motivated by the idea of doing something to help other people," she says simply.

OLYMPIA SNOWE

When I was still in Congress, a woman who had been a friend of my mother's sent me a letter. It turned out to be a letter that my mother had written to her. I was very touched to have it. She wrote of the day she went to the hospital. My father,

John, and I were standing in the pouring rain as my mother was getting ready to leave. My mother wrote that I kept telling her I would take care of her when she came back home. I'm sure I wanted to make things better. I know I tried. I remember once when my mother was in the hospital, I decided to help my father by doing the laundry. I dumped all the clothes in the bathtub, poured in laundry soap, washed them by hand, and then hung them on the line. My father wasn't too thrilled when he came home because I'm sure I never got all of that soap out of the clothes.

It was very hard for my father after my mother died. He was also not in good health, and he worried about how he would raise two children alone. And so my father made a very difficult sacrifice. He sent me to St. Basil's Academy. I know how wrenching it was for him to make that decision, but it ultimately had a profound influence on my life.

Olympia Snowe believes her reputation for independence can be traced directly back to her time at St. Basil's. By the age of eleven, she was taking the train back and forth from Maine to New York by herself. This meant a long ride from Maine to New York City's Grand Central Station and then a change of trains— and sometimes an overnight stay at the station—in order to reach her school. One morning, after she slept in the station while waiting for her train to Maine, her suitcases were stolen.

Sometimes Olympia was entrusted with the care of other children, helping to usher them to their trains before finding her way to her own.

OLYMPIA SNOWE

It was also at St. Basil's that I got involved in politics—and I loved it. During the eighth grade, I ran my first campaign for office—as dorm president. I had political posters and gave campaign speeches, standing in the dining room before my classmates. Best of all, I won!

Even when I was very young, I always had the thought in the back of my mind that I wanted to do something special with my life. I always believed in myself. It never occurred to me that I might not be able to do what I set my mind to. The strength of that belief was derived from the encouragement of my family.

When she finished the ninth grade, Olympia returned home to Auburn, Maine, to live full-time with her aunt Mary and uncle Jim and their family. The next year her uncle Jim died.

After her graduation from Edward Little High School in Auburn, Olympia went on to the University of Maine in Orono, where she majored in political science. As a college student during the turbulent 1960s, Olympia was certain of her Republican beliefs. But, even then, her bipartisan proclivities were evident. With the assistance of Charles O'Leary, who would later become president of the Maine AFL-CIO, she secured a summer position in the office of Governor Kenneth Curtis, a Democrat.

By the time she graduated from the University of Maine,

Olympia was engaged to Peter Snowe, a young businessman from Auburn who had previously served one term in the Maine House of Representatives. They married, and she plunged into Republican politics in her hometown, winning a seat on the Board of Voter Registration, hosting coffees, and working on the campaigns of statewide candidates in 1970 and 1972, including the final reelection campaign of Senator Margaret Chase Smith and the first congressional campaign of Bill Cohen, who would later go on to become a U.S. senator and secretary of defense.

In 1972, Peter Snowe again sought and won a seat in the Maine Legislature. He was sworn into office in January 1973. At the same time, newly elected Congressman Cohen had asked Olympia to help set up his office in Lewiston. It was while she was at work in Cohen's office that she received the devastating news that her young husband had been killed in an automobile accident returning home from the legislature in a snowstorm just three months into his term.

OLYMPIA SNOWE

In a fateful twist of irony, on the day Peter was sworn into office as a legislator, we took some photos on the House floor. We took pictures of Peter sitting in the chair at his desk and me standing beside him. Then he said to me, "Olympia, now you sit in the chair." I didn't think I should, but he insisted, so I sat at the desk and he stood next to me. That was in January. By May, I was sitting there in his place.

In the days after Peter Snowe's death, many friends, neighbors, and local leaders urged Olympia to consider becoming a candidate in the special election to fill Peter's seat in the Maine House of Representatives. The young widow, still reeling from this latest tragedy in her life, decided to take the challenge.

For Olympia Snowe, each of the losses she had suffered served as lessons in the realm of the possible. She quickly learned that given the qualities of character and tenacity, she could overcome every obstacle and make a difference. And along the way she also learned that, given her own experiences, she could make a difference to benefit others who faced similar crises. "I have always tried to take negatives and turn them into positives. With the devastation of Peter's death came a sensitivity to the tremendous difficulties that other women in similar situations can face—such as raising children alone. Later that was brought to bear on issues such as pension reform, child care, and displaced homemakers. And since I've served in a position that can make a difference, I have always worked to translate those lessons into a force of positive change for others."

Harriet Beecher Stowe once wrote, "When you get into a tight place and everything goes against you, till it seems as though you could not hang on a minute longer, never give up then, for that is just the place and time that the tide will turn."

Every woman who reaches the pinnacle of elected office does so by climbing over barriers. For some, the barriers are greater than for others, but they are always a key component of the climb. The price of triumph is often tragedy. Just as real as the price of winning is the possibility of losing.

said that public service is in her genes. Five generations on her father's side of the family have served in the Maine legislature, dating back to her great-great-grandfather. Both her father and mother served terms as mayor of Caribou. Her father served in both the house and senate of the Maine legislature. Her mother chaired the school board in Caribou and would later become chairman of the University of Maine System board of trustees. She would make her contribution to the community while raising six children, a living example of the values she taught.

In a bow to the powerful combination of nature and nurture, Susan Collins found herself drawn to public service at a young age. As a child, she was an avid reader, and she especially loved stories about the courageous, independent women who broke the mold and had an impact on the nation and the world in spite of overwhelming obstacles. She was a natural leader. In high school her classmates elected her president of the student council. In college she became the first student ever to chair the academic affairs committee, on which faculty, administrators, and students served.

While she was a student at St. Lawrence University in Canton, New York, Collins became impressed with a young man named Bill Cohen who was running for Congress in her home district in 1972. Cohen was a moderate Republican with an independent voice that Collins found refreshing in the turbulent political environment of the early 1970s. When she was home from school, she volunteered to work on his campaign. When Cohen decided to walk the length of the second congressional district to meet the people, Collins followed behind him in a car filled with supplies and campaign literature, as he walked the hills and valleys of Aroostook County. The citizens responded to Cohen's grass-roots appeal and elected him to Congress.

Collins interned in Cohen's Washington office in 1974 before beginning her senior year at SLU. What a time to be in Washington, D.C.! This was the summer of Watergate, and Collins found herself in the center of the maelstrom. When Cohen became one of the first Republicans to break ranks and call for President Nixon's impeachment, his congressional office became the focus of national attention. Collins was assigned to help field calls and open letters from constituents. It was a trial by fire, and she took away one important lesson: Keep true to your conscience and learn to accept the outrage as well as the praise. It was a lesson she would apply on many occasions during her own political life.

In 1975, after graduating magna cum laude from SLU, Collins was hired as a full-time staffer in Cohen's Washington office. She remained with him for the next twelve years as Cohen moved from Congress to the Senate. Collins spent six of those years as staff director of the Senate Subcommittee on Oversight of Government Management and found that she loved policy work. She discovered that she had a talent for figuring out rational, workable solutions to extraordinarily complicated problems. It is a valuable gift.

In 1987 Collins returned home and served as commissioner of professional and financial regulation in the cabinet of Maine governor John R. McKernan, Jr. She stayed for five years, then joined the federal government during the last year of the Bush presidency, taking over as the New England administrator of the

Small Business Administration. In each job she honed her skills as a leader and a problem solver.

Never one to blow her own horn, Collins was often underestimated by others in the beginning, but she invariably won them over by her dogged dedication and her bright mind. People who worked with her grew to respect her quiet ability to get to the bottom of a problem and devise a solution that pulled people together rather than drove them apart. Her concern was always genuine, and she expressed it in ways that mattered. She began to make a name for herself because she got results.

In 1993, when she was approached about running for governor of Maine, she was only forty years old and had never held elective office. But she took the risk because she knew she could do the job.

SUSAN COLLINS

Someone once asked me, "Whatever possessed you to run for governor as your first elected office?" It was a valid question. But by the time I ran I had been working in government for eighteen years, in a variety of policy and management roles. I'd been in Governor McKernan's cabinet for five years. I had run a state department with two hundred people, been in charge of financial regulation and licensing boards, and tackled some very tough issues, like workers' compensation and health care. So I knew state government very well, and had worked closely with the legislature. In addition, my executive experience had been strengthened by my time as the New England chief of the Small Business Administration.

I was a political appointee in the Bush administration, and my job ended on Inauguration Day, 1993. It just so happened that Governor McKernan would be finishing his second term at the end of 1993. Since he could not run for reelection, people were trying to figure out who was going to run for governor and take his place. I started getting calls and letters from people whom I'd worked with in the state government. They encouraged me to run, including many of the state's civil service employees who'd worked for me. I always took that as a great compliment, particularly since most of them were Democrats. But they believed that I'd be a good leader. They knew that I would respect them and work well with them.

I had a lot of respect and admiration for the other seven candidates in the Republican primary, most of whom I knew well. But I realized that I had something different to bring to the table—I had experience as an executive. And, more specifically, I had experience running a department in the state government, managing budgets and setting policy.

Still, I had never run for elective office before, and governor was a pretty big leap. But I remembered a speech I'd heard Lynn Martin, the former congresswoman and secretary of labor, make a few years earlier. Secretary Martin said that she'd found that women who would make very good elected officials all too often talked themselves out of running for office. She remarked that women too often think they're not prepared, tell themselves it's too soon, think that they should wait. I thought about her words, and I knew they were true. I absolutely believed that I could be a good governor and could make a real contribution. So I decided to go for it.

It was grueling, but Collins went on to win the eight-way Republican primary. She was energized by this vote of confidence but found that the general election was brutal. It was a very difficult campaign. Early on, a disgruntled extremist from Southern Maine filed a suit against Collins, claiming that she wasn't a resident of Maine and that she had given up her residency because she'd worked out of Boston while heading up the New England SBA. He demanded that Collins's name be removed from the ballot.

When Collins first heard about the charges, she laughed. Her family had been in Maine since 1760. She owned a house in Maine—the only house she'd ever owned. She'd voted in every Maine election. She was a Mainer through and through.

But she stopped laughing pretty soon. Although it was a frivolous suit, Collins still had to hire lawyers and spend time defending herself, during the crucial months when she should have been building her campaign. The extremist lost at every stage, but that didn't deter him. He kept appealing. The suit went all the way to the Maine Supreme Court, which ruled unanimously in Collins's favor.

The suit had drained time and energy at a critical time in the campaign, and that hurt Collins. "I'm more experienced now, and I'd know not to let that kind of diversionary tactic get in my way and interfere with a campaign today," Collins says. "But at the time, I did what I thought I had to do."

Collins lost the general election in the fall, coming in third behind the self-financed Independent candidate, Angus King, and former two-term governor Joseph Brennan. The Green Party candidate finished a distant fourth.

"I got clobbered," Collins says, laughing. "But despite the obstacles and the outcome, I learned so much from the experience and made wonderful new friends."

SUSAN COLLINS

I made plenty of mistakes in my first campaign. But in the end, I think it was money that made the most difference. It was my first experience seeing just how crucial money is in determining who gets elected—unfortunately. The seeds of my interest in campaign finance reform were sown in that election. For me, it was very simple: Were we reaching a point where people like myself, who loved public service, could not fulfill their vocation because they lacked huge war chests or were not personally wealthy?

Running for governor was an exhilarating experience, but it's very difficult to lose a race. You feel that you've let people down who believed in you and worked so hard for you. People have donated money, helped put together rallies, volunteered countless hours. I felt responsible. And then I also had a very practical problem. After campaigning full-time for a year, I was flat broke. My meager savings were gobbled up since I had been without income for a year. Like so many Americans, I found myself juggling bills and wondering if I could afford health insurance and still pay my mortgage. When the chance came for a job as executive director of the Center for Family Business at Husson College, I was delighted, and I started immediately. I knew the job at Husson would give me the opportunity to pursue two of the planks of my platform: helping small businesses and promoting education. The election took place in November, and I started my new job in December. I lived temporarily with a friend in Bangor and then moved to my own apartment once I could afford the security deposit. I started rebuilding my savings and getting on my feet financially again.

Losing an election can feel like a kick in the stomach. It is a very public rejection, and so is unlike any other. You put yourself out there and allow a bright light to be shined on you. The people collectively watch and listen. Given the choice, they say no. No matter what face is put on it, closing a campaign office after a defeat is a painful exercise, whether the candidate is a man or a woman. The question is, can a candidate lose an election and land on his or her feet?

The answer is a qualified yes. There is no question that it has always been more difficult if you're a woman. Women receive a more intense scrutiny than men, and the old observation of Clare Boothe Luce remains true: "Because I am a woman, I must make unusual efforts to succeed. If I fail, no one will say, 'She doesn't have what it takes.' They will say, 'Women don't have what it takes.'"

Still, things are changing. Today the political lesson accepted and understood by the women of the United States Senate is a lesson that men in politics have long known: in the quest for victory, one must always be willing to risk defeat. Passion for public service has to be greater than fear of losing a political race.

For Susan Collins, that passion didn't die. It was merely redirected. She loved her job at Husson. It allowed her to be involved in her two greatest interests—education and helping small businesses create jobs. Collins was happy and fulfilled. It had taken a year, but her finances were stable again. She'd even managed to put a down payment on a small house in Bangor.

Then Maine's senator William Cohen announced that he would not be running for reelection when his term ended. Susan Collins's phone started ringing off the hook.

SUSAN COLLINS

In two days I received more than one hundred calls. People from all over the state were encouraging me to run, including people who said that they hadn't voted for me the last time because they felt they didn't know enough about me. I'll never forget one woman telling me that I'd been such a gracious loser that she felt she owed me a vote.

Now what did I do? I was happy at my job, I was finally solvent, and I knew I was staring at a long stretch of no income—again—while I campaigned. Did I want to take that risk again? Could I put myself through all of that again? Could I risk losing again? It was so tempting to just say "No, the time isn't right." But I couldn't do it.

I kept imagining myself at eighty-five, sitting in my rocking chair before my fireplace, thinking about my life. I didn't want to look back and say, "If only I'd had the guts. If only I could do it again." I didn't want to have any "if only's" dogging me for the rest of my life. I knew I loved public service, and I wanted to serve the people of Maine. I didn't want to be one of those people who are filled with regret about not taking chances, not taking risks, not seizing opportunities. I knew too many people who were like that, who spent their lives bitterly disappointed that they didn't do what they really wanted to do. I was happy with my life, but I wanted more. I wanted to achieve my dream.

I'm glad I took the risk. But what I really learned wasn't just that the risk is worth it if you win. As long as you keep your head on straight and aren't afraid of failing, you gain something even if you lose. New doors open. It's not true that we only get one chance; we get as many chances as we're willing to take. I'm glad I decided to run. I can make a real difference here, and that means everything to me.

Two of the women senators, Kay Bailey Hutchison and Mary Landrieu, faced challenges to their honesty and integrity, which were far more difficult to overcome than the more familiar challenges of campaign rhetoric.

Kay Bailey Hutchison was a seasoned politician when she decided to run for the United States Senate in a 1993 special election against the incumbent senator, who had been appointed by the governor to fill the unexpired term of longtime and highly respected Democratic senator Lloyd Bentsen. Bentsen had been named secretary of the treasury in the Clinton administration.

Hutchison already knew how to accept defeat as well as victory—she had seen her fair share of both. She was usually able to deflect criticism and not take it too personally. As Texas's first Republican state treasurer in modern times, she was outspoken about some of the policies of the overwhelmingly Democratic state leadership. She cut her office budget and fought against a state income tax. She had long ago come to terms with the fact that not everyone appreciated her conservative policies and good intentions.

She felt strong and confident when, in June 1993, the people of Texas overwhelmingly elected her to the United States Senate as the first woman in the state's history to hold that office. But her eagerness to begin representing the nation's second-largest state was tempered by her awareness that she faced another election soon. Bentsen's term ended in 1994. Hutchison immediately began campaigning for reelection to her first full six-year term.

It was then that she was hit with a sucker punch.

At the very moment of reaching the pinnacle of her career, Hutchison was charged with venality. She had not seen it coming. It seemed that she might be toppled in the worst possible way, through an indictment on charges that she had used the state treasurer's office and its staff for political purposes.

"I knew it was political," she says today. "So did the people of Texas. I received overwhelming support from people all across the state, who told me that they knew it was political. But when you are indicted under the direction of political adversaries, saying you broke the law and abused the public trust, it doesn't really help much to know that it's not true and is politically motivated. It's a crushing blow."

The Democratic establishment in Texas was understandably shocked by Hutchison's victory. She had defeated an incumbent senator by the largest margin ever recorded in United States history. This was an unexpected drubbing for a party that had controlled the political destiny of the state since 1875, and which had held one or both of Texas's United States Senate seats since it became a state in 1845.

Viewed from this political perspective, the indictment's purpose seemed clear. With just one year remaining in the Bentsen term, the legal proceedings could derail Hutchison's campaign for reelection the following year. The specter of a high-profile trial pending for a sitting United States senator would be devastating to any senator's reelection efforts.

Hutchison demanded, and the court approved, an immediate and speedy trial—just one of many of the constitutional rights of all Americans that Hutchison learned firsthand to appreciate.

She was determined to make her way through the ordeal without breaking. In the coming weeks, with a trial hanging over her head, she focused on her work and kept her public demeanor strong and steady. There would be no angry attacks, no bitter recriminations. Just calm and focus.

As the trial date neared, talk of a plea bargain filled the political air. There were rumors that it might include an agreement that she would withdraw from the Senate race in 1994. Hutchison made it known that such an agreement would never be an option. Her attorney, Dick DeGuerin, sent word: "She's not pleading guilty to spitting on the street."

When the trial came and the jury was seated, the judge asked the DA to proceed with his case against Hutchison. In an astonishing development, the district attorney declined to do so, and asked the judge to dismiss the charges. The judge asked the DA if he intended to reindict Hutchison before the election; he said he wouldn't rule it out. The judge rejected this strategy. Instead of dismissing the charges, he instructed the jury to deliberate and return with a verdict. With no evidence, the verdict was not guilty. The case against Senator Kay Bailey Hutchison was over. She had won.

The cloud hanging over Hutchison lifted just as suddenly as it had appeared. She went on to win reelection to the United States Senate in 1994, defeating Texas investment banker Richard Fisher with 61 percent of the vote.

There is an adage: "That which does not kill you makes you stronger." Hutchison would have preferred another path to strength, but all who know her believe that she is stronger for it.

KAY BAILEY HUTCHISON

A friend of mine gave me a book after the whole thing was over. It was called *When the Worst Thing That Could Happen in Your Life Already Has.* The book talked about how experiences like mine could be liberating, and that's true. Before I went through the indictment, I used to worry about how I came across in a debate, the way I was portrayed in a newspaper story—the kinds of things that can nag at you when you're in the public eye. But after I went through that horrible period, and then I finally won, I just didn't worry about things the way I used to. It's given me a different perspective, a renewed respect for our Constitution and the Bill of Rights, and a deeply held sense of gratitude for the people of Texas who didn't buy into the scam.

Mary Landrieu's ordeal would take almost a year to resolve.

November 1996—New Orleans, Louisiana

It was a long night. As the votes streamed in from across the state, they never quite coalesced to form a picture of the winner of the race for the United States senator. It was going to be close.

It was well past midnight before Mary Landrieu was declared the winner—by a mere 5,788 votes.

A close race is always agonizing while the votes are being counted, but it turns glorious when you win—especially when the victory is carved out of a previous loss. Just a year earlier Landrieu had lost a spot in the runoff for governor by about the same number of votes. Forever after, it is the exquisite memory of the seconds preceding victory that is savored over and over again. In the same way, the winner in a tight political contest may experience a special thrill because the battle was so hard fought, but in the end she prevailed.

In addition, for Mary Landrieu, the win was especially meaningful, because her opponent, Louis "Woody" Jenkins, was a man who stood for many things Landrieu deplored. A case in point: during the campaign, it was discovered that the Jenkins campaign had purchased a mailing list from a company affiliated with the ex–Ku Klux Klan leader David Duke.

Landrieu had a long history with Jenkins, dating back to the time when they served together in the Louisiana Legislature. She found him to be an ideological obstructionist and a fabricator of plots——the kind of politician who would rather throw a wrench in the works than collaborate on important legislation.

So it didn't come as a surprise to Landrieu when Jenkins challenged the Senate election results, accusing Landrieu of widespread fraud, involving thousands of "phantom votes." However, she didn't take it seriously. She thought it was just a typical Woody Jenkins maneuver.

MARY LANDRIEU

I knew we'd done nothing wrong. And I wasn't terribly concerned because few people in the state took his accusations about me seriously. I just figured the members of the Senate Rules Committee would immediately see that there was no evidence and dismiss the matter. So, when they wouldn't dismiss it, and the months dragged on, I started to worry. I thought, "They're actually taking it seriously!"

The investigation dragged on. The General Accounting Office found no evidence of fraud. Still, it didn't end. By the time the investigation was concluded in my favor nearly a year after the election, it had cost me $750,000 in legal fees and expenses to keep the seat I had won.

It wasn't exactly the kind of welcome I was expecting in the

Senate, but I have no bitter feelings. And in some way, it was a benefit. As a freshman, I had a good opportunity to work with the leadership on both sides. To this day, Trent Lott and I have a special connection because of that experience. I just didn't let it beat me. One thing you learn early about politics is that this is a tough business, but worth it.

5

Balancing Acts

*Jim and I get up very early—about 6:00 A.M. We bathe
and dress the children and feed them a wonderful breakfast.
Then we put them in the freezer, leave for work, and when we
come home, we defrost them. And we all have a lovely dinner
together. They're great!*

FORMER CONGRESSWOMAN PAT SCHROEDER,
in 1972, in response to persistent press questions
about how she could serve in Congress and
take care of her children at the same time.

February 2000—Dirksen Senate Office Building

The reception area of Senator Blanche Lincoln's office is
crowded with a group of conventioneers from Arkansas who
dropped in to say hello. Can she spare a minute? This happens a
lot. Lincoln always looks genuinely pleased when a familiar face
from back home pops in. Unfortunately, time is limited—there's
never quite enough of it to do all the things she'd like to do.

Just as she rises from her desk to greet her visitors, a loud electronic bell sounds in the Dirksen Building, the signal that summons senators to the Capitol for a vote. Apologizing profusely, Lincoln grabs her bright red jacket, which is hung over a chair, and slips it on. Then, accompanied by a legislative aide, she hurries out the door and down the hallway. As they wait for the elevator, Lincoln absentmindedly rubs at an almost indiscernible stain on her jacket sleeve. "Peanut butter," she mumbles, shrugging helplessly.

At the basement level, a rotunda feeds onto a platform where a tram slides back and forth along a track that runs from the Capitol to the three Senate office buildings. Lincoln decides to walk, and they take off in a rapid stride down the walkway, arriving at the Capitol building ahead of the tram.

Lincoln is the mother of four-year-old twin boys, so she is accustomed to living life on the run. Ebullient, energetic, and unpretentious, she is the picture of representative government. "It's hard to be high-and-mighty when you've got peanut butter on your sleeve," she laughs.

Lincoln has some of the same easy Arkansas charm that President Clinton possesses. She loves people, and relishes campaigning in the rural communities. "The way you win in Arkansas is to be out there with the people," she says. "Eating barbecue, riding on flatbed trucks, and waving during parades." Lincoln's Senate Web site includes "The BBQ Bandwagon," devoted to listing the hottest barbecue spots in the state.

Lincoln seems like a natural politician, but it is not a career she ever anticipated. She wasn't really sure what she was going to do with her life. She grew up as part of what she calls the "in-between generation." She wasn't necessarily expected to stay home and not work, but neither was she presented with a lot of options. She thought of herself as a woman in a new generation, who didn't quite fit into the old puzzle.

In college, Lincoln studied biology, thinking she would pursue a career in the medical sciences. After graduation, feeling a little burned out from the heavy course load she had carried, she decided to take a year off before going on to graduate school. She took a job in Washington, as a staff assistant to Arkansas congressman Bill Alexander, because she had always been interested in government. After that, she worked in the private sector, handling government affairs. She stayed for nine years. Along the way, Lincoln discovered her true passion. It gradually dawned on her that she wanted to make her government better and more effective. Not only that, she wanted to do it for the people of Arkansas, her home state. It was the place she loved most in the world.

BLANCHE LINCOLN

I truly, truly loved my home, and I wanted to do something for the people there. I was having lunch with a friend in Washington one day, and I told her of my frustrations, and how I didn't feel fulfilled because I wanted to be doing something for my home. And she laughed and she said, "Well, you should run for Congress." I laughed, too, and then we pulled out a notepad and started writing down what I'd have to do. And we were saying, "How hard could it be to put together a list of your friends and family? How hard could it be to get in the car and ride around to twenty-five counties? What's twenty-five counties? It's not that impossible, is it?" Suddenly I realized that I could do it. Gosh, other people do it! So I went home and told my family, and it kind of threw everybody for a loop. I mean, I was a young, single woman. What was I thinking? Of course, I had two older sisters and a brother who pretty much paved the way. I don't think there was much I could do with my life to shock Mother and Daddy. Running for Congress was probably the best thing I could have done if I wanted to try to one-up my sisters. We talked it over in the family, and I told them, "I really think I can do this, and I know I want to." They said, "Then do it. We'll help."

Mother and Daddy taught us kids to take a sense of ownership in things. Daddy always said, "I don't want to hear you whining about things you're not willing to do something about." He'd tell us, "Either stop griping, or get out there and change it." And I felt frustrated with government. I think I was probably a prime example of the mood of the electorate. I felt like government was being far too reactionary and not as progressive and productive as it should be. So I took Daddy's advice and did something about it.

There was only one twist to the plan. The congressional seat Lincoln would be seeking belonged to her old boss, Bill Alexander, now a twenty-four-year incumbent. Lincoln went to see

Alexander and told him she was running against him in the Democratic primary. He smiled and wished her luck. Obviously, he didn't take the challenge seriously.

Lincoln returned home to Helena, Arkansas, and moved back in with her parents. She set up an office in their house. Her mother enlisted her friends from the church to help stuff envelopes and make phone calls while Lincoln traveled during the day and cranked out literature on her computer at night.

BLANCHE LINCOLN

One of the first things we had to do was hold a press conference to announce my candidacy. We'd never organized a press conference before, but Mother's friends from the church offered to make food. They figured it was an event, and in the South when you have an event, you feed people. So there were big pots of chili, and brownies and hot dogs, and balloons for the kids. When the press arrived, their jaws dropped. They said, "What's *this?*" They'd never been to a political press conference where they got *fed*.

I put some miles on my car running for Congress. I traveled all across east Arkansas. The district was twenty-five counties, but the largest city had only about sixty thousand people. So it was very rural, and that presents a challenge if you're going to reach the grass roots. You've got to be out there in the coffee shops, the farms, the lunch counters at noon. I won the election and went to Washington the same year Clinton was first elected. I served for two terms of two years each. During that time, friends in Arkansas introduced me to Steve, who is a surgeon, and we got married while I was in Congress. When I became pregnant and then discovered we were going to have twins, I made a choice not to run for reelection, although I did finish my term. My kids were due in early July, and I just didn't see myself riding around on flatbed trucks being eight months pregnant with twins, and seventy-five extra pounds out to there.

My husband and I wanted to do all that we could to ensure that it would be a safe and healthy delivery. I was thirty-five years old—young for Congress, but old for motherhood. The doctors made me feel like Methuselah. I think the term they used was "maternally challenged." So I didn't run for reelection, but I finished my term of office after the boys were born. Then I stayed home for a year, until Senator Dale Bumpers announced that he was not going to run for reelection. A lot of people came to me and said, "We hope you're interested. Do you think you could do it?" So Steve and I sat down with our legal pads, and we listed the pros and cons. The big question was, Could we go to Washington and make it a productive experience for our family? Could we make a contribution to our state, and still raise our family and do the things that were important in our lives?

We talked about it, and we decided that we could really do it, and do it well. Steve could join a medical practice in the Washington area. The entire effort was dependent on the partnership I have with my husband. He's been very supportive and very helpful. I absolutely couldn't have done it without him being 100 percent behind me. But, more than that, it wouldn't have been any fun without him. Steve is my best friend.

Lincoln follows Hattie Caraway as only the second woman ever to be elected to the United States Senate from Arkansas. Her predecessor could not possibly have imagined a time when a young mother would be filling her shoes. Lincoln's days are a juggling act that all working mothers can identify with. She is up at the crack of dawn to get the family ready, feed her sons, Reece and Bennett, and then bundle them up and take them to preschool. Sticky hugs and kisses and reassurances trail her as she leaves: "Mommy's going to work. I love you."

If anything frustrates Lincoln, it's the near impossibility of planning her schedule, of never being able to anticipate what each day will bring.

BLANCHE LINCOLN

It just drives my husband crazy. Steve's a total type A. As a surgeon, he knows precisely when his surgeries are scheduled. He knows how many patients he's going to see every day, and when he's going to see them. Sure, he has emergencies, but for the most part he's organized. It amazes him that I can't schedule my days that way. When I complain that I have trouble clearing enough time to spend with my staff, Steve says, "Well, you just need to decide that every day at eleven o'clock, you're going to sit down with them." My response? "Yeah, right. And what happens if five constituents show up and that's the only time they can meet with me? Or someone calls a caucus meeting? Or we have a vote? Or there's a committee hearing that I have to attend?" It would probably surprise a lot of people to know that in the Senate, you can't mold and shape your schedule as you like, because you don't have a lot of control. It was different in the House because there was a Rules Committee, and everything was pretty rigorously scheduled. But I think our forefathers envisioned that the Senate would have greater flexibility, so that it would be easier for the senators to talk about and explore the issues with one another, look for consensus, take time to get to the bottom of things. So I've gained flexibility being in the Senate, but I've lost the predictability I had in the House.

Sure, our life is hectic, but that's no different than the way it is for most working families today. I certainly have a better appreciation of what kinds of pressures they face. I tell them— "I'm just like you." It's something I try to impress on the children when I speak at elementary schools. I want them to know that government really is for the people, by the people. It's funny because whenever I go to a school and I open it up for questions, you know what they ask? They ask, "Do you ride in a limousine? Do you have a bodyguard?" They envision this life in the lap of luxury, where I'm driven around in a limo, and somebody's doing my nails and fixing my hair. And I'm having lobster and champagne for lunch. I tell them it's more like Diet Coke and turkey sandwiches. And limo? I wish. Usually, I'm begging one of my staff for a ride home. I try to leave the impression with them that I'm not any different than they are—which is another way of saying, "You could grow up to be a senator, too."

Before 1992, only one married woman had ever been elected to the United States Senate. And before 1992, when Patty Murray and Carol Moseley-Braun were elected, a woman with children still living at home had never been sent to the Senate. But with the elections of Mary Landrieu in 1996 and Blanche Lincoln in 1998, it's no longer unusual to come across preschoolers racing down the hallways of the Dirksen and Hart Senate office buildings.

When Lincoln arrived at the Senate, she found a strong ally in Mary Landrieu, who was already combining the roles of mother and senator.

She'd met her husband, Frank Snellings, eleven years earlier, during her campaign for state treasurer. Snellings was the president of the Ouachita parish. (In Louisiana, parishes are comparable to counties.)

"My area campaign manager introduced me to Frank," Landrieu recalls with a laugh. "Now, *that's* a campaign manager! He helped me get elected, *and* he helped me get married." They went out on a date and were instantly attracted. Within seven weeks they were engaged, and their whirlwind courtship culminated in marriage six months later. "I was surprised at myself," Landrieu admits. "I'm not the type of person who rushes into things. But I felt great peace about my decision. I fell in love, and it was right."

Frank Snellings was the kind of man Landrieu had always hoped to marry—a man who would see her as an equal partner, and would share her goals. He was comfortable with her line of work because his mother had been a very popular elected official when he was growing up. "I'm strong-willed, but I'm a piece of cake compared to Frank's mother," Landrieu jokes.

After several years the couple adopted a baby boy, Connor, who was four years old when Landrieu was elected to the Senate. Then, during her first year in Washington, a baby girl became available for adoption. It was right in the middle of the investigation of Landrieu's campaign, and might have seemed like the worst possible time. But Landrieu was determined that the adoption go through. "I'm sure my colleagues thought, 'Now we *know* she's crazy,'" Landrieu says. "But Frank and I weren't about to let that fraudulent investigation derail our life and rob us of this great joy. So we adopted Mary Shannon."

She was delighted when Blanche Lincoln arrived with her twins; now there were two of them to serve as role models for young mothers. She fiercely defended Lincoln against critics. In one notable instance, the conservative columnist Mona Charen wrote a piece entitled "Return from the Nursery?" lambasting Lincoln for being an irresponsible mother. Charen wrote, "Running for the U.S. Senate is incompatible with being a good mother to very young children."

Landrieu was indignant. She sat down and wrote a response, which was published under the title "Mothers Make Good Senators, Too." It was a clear and compelling articulation of the precise reason why it was so important to have mothers in the Senate. She wrote:

> When I was a young woman growing up in New Orleans, the daughter of a controversial and inspirational mayor, many things were said and published about my dad. Stories and columns were normally good, but as with any leader some were negative, too. However, no critic ever questioned Dad's capability

of being a good father to his nine children simply because he was mayor of New Orleans.

Throughout much of 1996 I ran for the Senate while being a full partner with my husband in raising our young son, Connor, and a year ago we were blessed with our second child, Mary Shannon. Partnership and teamwork are the keys for any working parents—running for and serving in the Senate are no exception.

Some 17 million women were employed in 1948, representing 28.5 percent of all workers in the country at that time. Today, because of economic demands, more than 60 million women work, representing 50 percent of America's workers. And millions of them are mothers with young children.

I have news for Ms. Charen. Mothers have been working full-time and raising successful well-adjusted children for decades. During the Great Depression, millions of women were forced into the labor market to help their families make ends meet. I would like to draw your attention to one courageous mother named Nellie. Nellie was married to an alcoholic dreamer. He was a well-meaning man who drifted from job to job and town to town, leaving Nellie with the primary responsibility of raising their two young boys.

To pay the bills and keep food on the table, Nellie took a job as a seamstress-clerk at a dress shop for fourteen dollars a week, and altered other people's clothing in the evenings and on weekends.

No one ever challenged her desire to work and make things better for her family. No one ever charged that because she worked twelve hours or more every day that she was incapable of being a good mother, even though she raised her children virtually alone.

To the contrary, Nellie's example of hard work served as an inspiration to her two boys and helped shape the way they viewed the world around them. They came to view hard work not as a liability but as an asset. Watching their mother work did affect their lives, but in a positive way.

It is the millions of women like Nellie who serve as my daily inspiration as I represent my state on the floor of the U.S. Senate. It's the women and men who balance incredibly tight work schedules with the complexities of rearing children day in and day out, who give me the hope that I can be a good senator and a good mom.

I have no doubt that my work as a United States senator, much like my father's work as mayor, will engender and not endanger my children. By the way, Nellie's two children grew up just fine and did pretty well for themselves, despite having a working mom. You may have heard of one of her boys, Ronald Reagan. He grew up to be the fortieth president of the United States.

When Patty Murray was elected to the Senate in 1992, her son, Randy, was fifteen and her daughter, Sara, was twelve. The family decided to uproot themselves and move to the Washington, D.C., area. Murray's husband, Rob, would find a new job, and the kids would change schools.

Sara was perhaps the happiest about the move. Murray's campaign had proven very difficult for her. What twelve-year-old wouldn't cringe at having her mom all over the newspapers and TV, posters of her everywhere? The attention embarrassed Sara, and it changed her relationships with her friends and classmates.

She had to deal with kids whose parents were opposed to Murray, and hear them echoing their parents' harsh judgments. One day she came home from school in tears. Some of the kids were saying that her mom was going to do away with their parents' jobs. She was glad when she got to move to Washington, D.C., and go to a new school where nobody knew her.

Five years later, when she was a senior in high school, Sara wrote an essay for her high-school newspaper, describing her frustrations at age twelve. By that time, she had adjusted to being the daughter of a senator, but she could vividly remember her deep feelings during the campaign. She wanted to be just Sara.

"Why should I support someone who was dragging me through hell so she could have a job?" she wrote. "No one ever cared to answer this trivial question except to say, 'She needs to do it so other women can.' This didn't answer my question. Who was my mother to suddenly be a spokesperson for the entire female sex?"

PATTY MURRAY

My kids have always been very outspoken, and it wasn't always easy for them. Randy is twenty-three now, and Sara is twenty. Sara's very passionate about issues. She's more of a feminist now than I am, and she speaks up for herself. Neither of my children has ever hesitated to disagree with the positions I've taken. When I was in the state senate, I worked on a bill requiring bicycle helmets for kids under fifteen. Randy was twelve at the time, and he hated wearing a bicycle helmet. He actually organized a petition drive against me at school. Then, during my first year in Washington, when Sara was thirteen, the same thing happened with her. There was a vote on trade. It was extremely difficult for me, but I made my decision, and I took a position that was unpopular with labor. Sara wrote a stinging editorial against my position in the school newspaper. But then she watched me calling our labor leaders in Washington State at home one night, explaining my position. They were painful calls, and some of the people were friends. Later Sara came to me and said, "I may have disagreed with you, but I really admire you for sticking up for what you think is right."

We always talk about how women senators don't have many role models. Well, Sara didn't have any role models for being the daughter of a mom who was a senator, either. I'm very proud of the way she worked through it.

Over the years, many people criticized me because I wasn't home with the kids. And they criticized me because my husband "had to" move to Washington when I got elected. What kind of mother was I? What kind of wife was I? But I got into politics because I'm a wife and mother. Who knows better about the issues that affect children and families than someone who is experiencing them?

Although Murray was frequently teased back in the Washington State Senate for how many times she mentioned her children in speeches, her philosophy of representative government resonates with the voters. All across America, women manage to work, raise their families, get their children to school, and maintain relationships with their partners. Why can't a woman who happens to have found work as a member of the United States Senate do the same? Murray believes that this is the core of her credibility.

PATTY MURRAY

The one thing I don't want to lose is who I am. I want to stay close to my roots. If I start forgetting, my husband always reminds me. The night before I was elected to the state senate, back in 1988, I was exhausted. I had been working like hell. I'd been doorbelling and licking envelopes every single day for ten solid months. It was really intense. So Rob said, "I'm going to take you out to dinner." I said, "How nice!" I felt like being pampered a little bit. So we got in the car, and Rob drove to Denny's. I said,

"What's this? You're taking me to Denny's?" He said yes, it wasn't a joke. I was too tired to argue. We went in and ordered dinner, and then Rob said, "I want you to look around this restaurant. These are the people that you'll serve as a state senator. These are the people you must remember."

I'll never forget that. It was an important moment for me. You can get carried away with an election and what it does to you—you can start thinking you're a member of some elite group. You need to stay grounded.

"We're just like you" is a positive message, and it's been a strong selling point for women politicians in recent years. That in itself represents a seismic change from the days when women like Pat Schroeder were forced to divert attention from their gender. Barbara Boxer remembers that in 1971, when she ran in her first election for county supervisor, "You never mentioned being a woman. You never brought it up. You hoped nobody noticed."

She laughs when she says this, because, of course, *everybody* noticed—and sometimes it seemed as if they noticed little else. Her children were seven and five at the time, but Boxer wouldn't have dreamed of running as a "mom in tennis shoes," as Murray did twenty years later. "If you were married, they thought you were neglecting your family; if you were single, they thought something was wrong with you; and if you were divorced, they were scared of you," she says.

It may be more accepted today, but the balancing act—being a committed public servant and just like everyone else—can be much more difficult when a husband and children are involved.

The Senate Wives' Club (now called the Spouses of the Senate) certainly isn't what it used to be. In fact, women in the Senate are often heard to remark that they wish *they* had wives. "Four years ago, we decided to move the family back to Washington State," Murray says. "When I go home, and there's nothing in the refrigerator, guess what happens? I head for the supermarket. Nobody else is going to do it. Let's face it. No matter how supportive, your husband is not usually the one who remembers to pick up the toilet paper."

My husband thought he married Debbie Reynolds and he woke up
with Eleanor Roosevelt.

BARBARA BOXER

There is one point that the married women senators can agree upon unanimously: it takes a special man to be married to a political woman. Kay Bailey Hutchison, reflecting on the encouragement, support, and advice that is generously supplied by her husband, Ray, says, "I have sometimes wondered if I could do this if I were married to a man who was not completely supportive of my choice, and the answer is no. Men like Ray don't get enough credit for all of the sacrifices they make."

KAY BAILEY HUTCHISON

My husband has a great sense of humor. When people ask him if it's hard being married to a woman who really enjoys talking about public policy, he just laughs and says, "No, I understand. It's a disease." He means the disease of government, and he's right. When I want to recharge myself after a grueling week in Washington, I'll go home to Texas and meet with constituents. And my bedtime reading? Let's see, right now I'm working my way through the budgets.

When Ray and I met, we were both members of the Texas Legislature. We were also on the ground floor of building the Republican Party in a state that had always been Democratic. In 1978, the year we got married, Ray ran for governor. It was a very tough election, and Ray lost. In 1981, I ran for Congress, and lost. So we were still very involved in the party, but we got out of politics for a while. Ray returned to practicing law. I went into business. I chartered a bank with a group of friends, invested in a designer showroom in the Dallas Decorative Center, and bought a candy manufacturing company with nationwide distribution. I loved being an entrepreneur. I remember at one point Ray and I talked about it, and we decided that I enjoyed the business world so much that if one of us went back into politics, he would be the one. But in 1990 we felt the tide in the state was turning for Republicans, and I had the chance to run for state treasurer. I looked at it and I thought, "This is the time, if I ever want to be in politics again." Ray was behind me 100 percent. So I ran and won.

The wonderful thing about Ray, in addition to his tolerance of my schedule, which means we spend much less time together than we'd like, is that he is completely happy with my success. He has no mixed feelings about whether he should be the one in public office instead of me. He's a rare man.

For Olympia Snowe, a different and truly unique sort of balancing act began when she remarried sixteen years after losing her first husband. She had known John R. "Jock" McKernan Jr. since they served together in the Maine Legislature. Later they were colleagues representing Maine's two congressional districts in the United States House of Representatives before he went on to become governor of Maine in 1987. They quietly dated, and Olympia adored Jock's young son, Peter.

McKernan has said he didn't propose to Snowe until he was sure he would get a positive response. "It took me a while to convince Olympia that the status quo was becoming unbearable. We needed some normalcy in our lives," says McKernan. On February 24, 1989, the couple married.

OLYMPIA SNOWE

Jock and I are soulmates—personally and politically. We have always been amused that when I have run for office, people have speculated that he was calling the shots, and when he has run for office, people have speculated that I was calling the shots. In fact, we both have always offered advice to each other. I guess you could say that I have been his closest adviser and he has been mine. But both of us ultimately make up our own minds about the issues facing us.

Jock has excellent political instincts, which I am sure he inherited from his mother, who served on the Bangor (Maine) City Council. One of the reasons he has always been so supportive of me, and of my passion for women's issues, is that he watched his mother work after the death of his father to balance the responsibilities of raising two sons and running the family's weekly newspaper.

Needless to say, given his background as a state legislator, congressman, and governor, Jock has a very real understanding of the demands I face as a senator. In fact, we often joke that our idea of quality time together is listening to each other's speeches. But in all seriousness, I know how lucky I am. I could not ask for a more supportive partner in life.

Combining a marriage partnership with a political partnership can be difficult—especially when you're doing it from different cities. McKernan and Snowe took it in stride. They were veterans of the life. A friend of Snowe's once observed, "Jock and Olympia have a closer relationship over the phone than most couples have seeing each other every day."

At the time of their wedding, Olympia was in her fifth term in Congress and Jock was serving his first term as governor. Their marriage marked another first for Olympia: She became the first person in American history to serve simultaneously as a member of Congress and a first lady of a state. At the same time she relished her role as Peter's stepmother, she also worked hard to balance the responsibilities of her job in Washington with the demands of her position as first lady and the rigors of tough re-election campaigns they both faced in 1990. After a grueling race, both Olympia and Jock overcame strong anti-incumbency trends that swept others to defeat in New England and were returned to office. Just days after Olympia and Peter stood proudly at Jock's side as he was inaugurated for his second term as governor in January 1991, tragedy struck yet again. Peter, by now twenty, a sophomore and accomplished athlete at Dartmouth College, collapsed during baseball practice, stricken with a previously undetected heart problem. Family and friends held a nine-day vigil at Peter's bedside before he died. Olympia has described the loss of their beloved Peter, so full of promise, as the most profoundly devastating of the losses she has experienced.

In 1994, barred by Maine's constitution from seeking a third term, Jock entered his second and final term as governor. Olympia, in her sixteenth year in Congress, had become the longest-serving Republican woman in the House. She was preparing to seek election to a ninth term when Senate Majority Leader George Mitchell, one of Maine's two United States senators, unexpectedly announced his decision to retire—leaving his seat open.

Olympia quickly announced her decision to vacate her House seat to seek election to the Senate, and the Republican field cleared so that she could run uncontested in the primary election. In the general election, she faced Democrat Tom Andrews, who held Maine's other seat in the U.S. House, as well as an independent candidate. Due to his wide election margins in the previous two House races, Andrews was initially viewed by many pundits as the favorite. Undaunted, Olympia waged a strong and effective campaign, pulled ahead early, and never looked back. In November, she became the first Republican to win this Senate seat since 1952, capturing more than 60 percent of the vote in a three-way race. She would be the only new woman elected to the Senate that year.

"It takes a very secure man to stand beside and support his wife when she runs for public office," Barbara Boxer says. "He has to know who he is and have confidence in himself. As a matter of fact, I'd say that today it's the *real men* who vote for women and support women in office."

Boxer has nothing but admiration for her husband, Stewart, and the seamless manner in which he has accommodated himself to her own personal growth. "I assure you, this was not something he bargained for. When we got married in 1962, we were looking ahead to living the American dream. But it was a fifties version of that dream. This was the year before Betty Friedan wrote *The Feminine Mystique* and woke women to the fact that we could follow our potential as far as it carried us. Stew was always on my side. He even talked me out of quitting when I was tired and discouraged. In his quiet, supportive way, he has made it possible for me to open doors for others. His contribution has been essential. It has also been historic."

When Dianne Feinstein was catapulted into the role of mayor of San Francisco, she had just buried her husband, Bert Feinstein, after a long illness. He had suffered from cancer, and Feinstein was by his side every possible moment until the end, working around the clock to perform her job on the board of supervisors and care for her husband. His death was devastating to her, and the memory of his excruciating pain has stayed with her, making her an advocate for reforms in the delivery of prescription painkillers.

Feinstein married Richard Blum, an investment banker, in 1980, and he opened her to new experiences. They went hiking in the Himalayas, they visited China, and they shared an interest in making a contribution to the world. It has been a solid, supportive relationship. However, Feinstein has no illusions about the role of a male spouse in a woman's political life. It's different for women.

DIANNE FEINSTEIN

If you are a woman who is meant to do this, you have to know that it will be a solitary road, and often lonely. People think nothing of it when a male politician walks into a room holding his wife's hand. In fact, they expect a show of solidarity from a political wife. But if I walk into a room holding my husband's hand, it's looked upon quite differently. People immediately wonder how much influence he has. They'll say my husband's the power behind the throne. After too much of that kind of press, my husband pulled back. Now I choose to campaign alone. That's okay with me. Historically, our role models have always been strong, solitary women. It's what is meant to be.

Women have always been the glue that holds families and extended families together. Today's woman is part of the "Sandwich Generation." She is not only the caretaker of children, but she is often called upon to be the caretaker of aging parents as well. Increasingly, midlife women have become unpaid workers in the role of providing full-time home health care to infirm parents. The socioeconomic strain on families can be enormous, depleting both their savings and their emotional resources.

Barbara Mikulski was elected to the Senate just as her father, Willie, was in the final stages of Alzheimer's disease. It was a cruel conclusion to a richly lived life. His death was devastating to the entire family, but especially to Barbara's mother. The feisty Miss Chris was a neighborhood legend. Her energy and spunk were in evidence every day on the streets of Highlandtown. She often worked in Mikulski's Baltimore office, and was a fitting surrogate for her daughter. It was plain to everyone that Barbara had inherited her lively personality and her gift for relationships from her mother.

But after Willie's death, Miss Chris's health began to deteriorate. Within two years, she was so ill that she could no longer go out. She suffered from diabetic neuropathy, and was often confined to a wheelchair. Mikulski and her sisters ensured her caretaking. It was a family effort with the senator and her two sisters each playing a role. One sister handled doctor's appointments, another bought groceries. Mikulski, because she lived only one neighborhood away, could be there quickly in the event of a nighttime emergency. She spoke to her mother every evening and helped care for her on the weekends.

Mikulski and her mother had a tradition of going out for lunch every Saturday. When Miss Chris became housebound, Mikulski brought lunch in, calling first to find out which of their favorite neighborhood restaurants her mother preferred.

With the family pitching in, Miss Chris was able to remain in her own home and still feel a part of her community. Mikulski feels lucky that she and her sisters lived near one another and could form a network of support for their parents. Still, it wasn't easy, and it sharpened her awareness of the struggle so many families were facing. It strengthened her commitment to making certain that Congress addressed the problems of hardworking, loving families who were striving to provide elderly parents with the dignified final years they so much deserved.

Mikulski had started fighting for seniors as soon as she arrived in the Senate. Her Spousal Anti-Impoverishment Law was passed in 1989, one year after Willie Mikulski died. Its goal was to protect elderly people and their families from poverty, in the face of steep medical expenses and the costs of long-term care.

"Honor thy father and thy mother is not only a good commandment to live by, but a good commandment for Congress to govern by," Mikulski said at the time. She would repeat the message ten years later when she launched a legislative initiative to provide a safety net for America's seniors. Her passion for the issue was grounded in her fierce devotion to her own parents, and her experience caring for them during their final years.

Miss Chris died on April 1, 1996. In her heartfelt, affectionate eulogy, Mikulski recalled the days when her parents Willie and Miss Chris, ran their grocery store and helped so many people through hard times:

> She and my father were a fantastic team running that grocery store. And as you know, they brought a tremendous amount of goodness and generosity into this community. When our grocery store closed in the 1970s, my mother volunteered in my office. That's where she got the great name—Miss Chris, the First Lady of Highlandtown.
>
> She helped run my neighborhood office. She

worked with me in the city hall when I served on the city council. Then she worked in my congressional office on Eastern Avenue, and then my Senate office on Highland Avenue.

She ran my neighborhood office. Whenever people would call, she would say, "Hi, I'm Barb's mother. What can I do to help you?"

If anyone asked where I was, my mother would say, "Don't worry about it. I'll take care of it. I'll tell her tonight, because I talk to her every day." And she did.

For all of us, my sisters, her grandchildren, she always loved us. She would leave us little messages on the answering machine. She would leave us little notes. She would also send her prayers. Because she believed that for every problem, there wasn't always a solution, but for every problem there was always a prayer that helped us get to the solution.

August 1999—San Francisco

Dianne Feinstein is enjoying an afternoon of rare pleasure—time spent with her seven-year-old granddaughter, Eileen. The two sit side by side in Feinstein's sunny pastel-yellow kitchen, drawing flowers with colored pencils on heavy white art paper. A vase of fragrant pink roses from Feinstein's garden sits in the center of the table as a model. Feinstein has some talent as a botanical artist. Her framed drawings line one wall in her Washington office.

Her California constituents know Feinstein as an authentic blend of motherly warmth and executive steel, but the first image that comes to mind is not grandma—or, as her granddaughter calls her, "Gagi." She would tell you that's exactly the point. Today's grandmothers are more likely to be out changing the world than sitting home in rocking chairs.

Feinstein and her granddaughter recently had the pleasure of collaborating on a children's book, in the "Grandmothers at Work" series published by the Millbrook Press in Connecticut. *Meet My Grandmother: She's a United States Senator* is full of color pictures of Feinstein and her granddaughter in settings that are both personal and governmental. Written from Eileen's perspective, it's a fun way for children to learn about how the Senate works and what it does.

DIANNE FEINSTEIN

This book is fantastic. We had so much fun doing it. The message that a grandmother can be a United States senator—well, as they say, I'm not your father's grandmother! On a more personal level, it will always serve as a wonderful reminder of this time in my granddaughter's life and of our times together. What grandmother wouldn't be happy and proud? Grandma's a senator! Pretty good role model, isn't it?

6

Different but Important

I didn't want to be one of the boys.
I did want to be one of the gang.

Barbara Mikulski

January 1993–Washington, D.C.

The old Senate chamber is a grand reminder of our young nation's architectural vision of the federal government. The detail was no doubt inspired by some of the great buildings and halls found on the European continent. The intricate ceiling pieces and columns, restored to their mid-nineteenth-century appearance in 1976, recall the work of another time and place. It is a jewel of a legislative chamber, semicircular and half-domed, located just north of the Rotunda. It was occupied by the Senate between 1810 and 1859. The Supreme Court then moved in and made it the judiciary's chamber from 1860 well into the next century, not leaving until 1935.

The first meeting of the Democratic senators was held in this chamber after the 1992 election, and the mood was upbeat and optimistic. The Democrats had regained the White House with the election of Bill Clinton to the presidency. The Senate had maintained its majority. And there was another victory, too. Its import was strikingly evident as Senate Majority Leader George Mitchell read the roll: "Barbara Boxer . . . Carol Moseley-Braun . . . Dianne Feinstein . . . Patty Murray."

As Barbara Mikulski sat beside her colleagues that day, her heart swelled with pride. Hearing those names called—*Carol, Barbara, Dianne, Patty*—she felt a surge of elation. She was no longer the only Democratic woman in the Senate. "I felt my shoulders relax," she would later explain. "The weight I'd been carrying seemed to ease." She made it clear that it wasn't because she'd felt isolated in the Senate. She had close relationships with a number of her male colleagues, and she knew that they supported her efforts. "I was by myself, but I was never alone," she said. Even so, there had always been within her a sense of her unique position, her representative oneness. She realized that she didn't just represent the people of the state of Maryland. Her second constituency naturally consisted of all the other women across America. Now there would be four more strong women to share the load.

Glancing up at the balcony, tastefully appointed when it had served as the gallery for female visitors so many years ago, Mikulski allowed herself to bask in a moment's reflective satisfaction. But only for a moment. Now that women were finally seated downstairs, there was a lot of work to be done.

When Barbara Mikulski was first elected to the United States Senate in 1986, the aura of a men's club was still so heavy that there may as well have been a sign hanging above the door. Mikulski was the first Democratic woman senator ever elected in her own right, and she was cut from a different cloth than the only other woman senator serving at that time, Kansas Republican Nancy Landon Kassebaum.

"My opponent in the election had argued that I didn't have the grace and polish to be a United States senator," Mikulski recalls. With a twinkle in her eye, she adds, "In other words, I didn't look the part."

But if some people expected Mikulski to be a Fells Point rabble-rouser, they soon learned otherwise. The real secret to her success quickly became clear—she was a masterful strategist. Mikulski had worked hard to get elected to the Senate, and she was determined to make a real difference.

BARBARA MIKULSKI

I didn't want to be one of the boys. I *did* want to be one of the gang. So I did two things. First, I asked for help. When I got to the Senate, I established a relationship with the old guard, starting with Senator Robert Byrd. I asked for his advice. I sought the advice of Senator Paul Sarbanes, who was the senior senator from my state. I was respectful of them, and respected the rules of the Senate. I went to all of the hearings and showed up on time. I read all of the reports and did my homework. I traveled with the other senators on fact-finding trips, worked hard on issues, became known as reliable. And because I did my homework and formed alliances, I got two important committee assignments—Appropriations, and Education and Labor.

I was still aware of my position, however. I was at an initial disadvantage as a woman coming to the Senate, and it wasn't just that the gym was off-limits. I didn't come to politics by the traditional male route, being in a nice law firm or belonging to the right clubs. Like most of the women I've known in politics, I got involved because I saw a community need. And it was tough, absolutely. I didn't have any natural mentors to show me the ropes. I had to seek out my mentors. So when four women finally joined me in the Senate in 1993, I was very gratified. I gladly took on the role of mentor and adviser.

———————————————————————————————

Barbara Boxer, the Democrat from California, had worked alongside Mikulski in the House of Representatives, and she trusted her insights. In 1989, when Boxer was considering a run for the Senate, she asked Mikulski what she thought of the idea.

Mikulski didn't mince words. "How old are you, Babs?" she asked.

"Almost fifty," Boxer said.

"Go for it," Mikulski urged her—but attached a caveat. "If you're ready to leave the House of Representatives, and never look back, and never regret it, then I'd say fifty is the perfect time," Mikulski told her. She added bluntly, "You can do more here, you can be heard here, and it's worth the fight you'll have to wage to get here. And it will be a fight."

Boxer nodded. She knew. It was one thing to have been elected to represent 500,000 people. It was another thing to seek election to represent 30 million people. If she decided to run, it would be an uphill battle all the way. Just getting herself that widely known would be difficult.

Mikulski gave her diminutive California colleague a reassuring wink. "Of course, you could do it for me, Barbara. I need someone around here that I can see eye to eye with."

"You'll never be one of the boys. And you should never want to be. Look at the Senate as a club with only a hundred members. Be collegial. Be a good sport. But be yourself. Don't try to model yourself after some concept of what a senator is 'supposed' to be. . . . You'll be different but important. Not one of the boys, but it'll help you be one of the gang. It's really quite logical."

SENATOR HILDA MENDELSSOHN's advice to a new senator, NORIE GORZACK, in *Capitol Offense*, a novel by BARBARA MIKULSKI and MARYLOUISE OATES (Penguin Books, 1996)

The Mikulski-Oates thriller series—the second book, *Capitol Venture*, was published in 1997—features Eleanor (Norie) Gorzack as the new senator from Pennsylvania. Norie Gorzack has a penchant for stumbling over cold corpses and chilling plots. Interestingly, there's a lot more to the books than murder and mayhem. They also serve as a noteworthy guide to any new senator facing the challenges involved—minus the dead bodies—in mastering the Mandarin-like influence structure of the Senate. Most notably, Mikulski gives Norie a wonderful mentor in Senator Mendelssohn, and it's no accident that the experienced Hilda's advice often echoes the very same advice that Mikulski has given to new senators.

Within a week of the 1992 election, Mikulski was already outlining an action plan for her new colleagues. She called it "Getting Started in the Senate." When the women arrived in Washington, Mikulski invited them to her office and conducted two separate seminars, during which she explained the complex workings of the Senate. In addition, she compiled thick notebooks for each of them containing all of the key details concerning committee assignments and procedures. There had never been any similar orientation training provided to new senators, and certainly there had never been a helpful information manual provided. The Senate had a long tradition of "every man for himself." Mikulski was determined that it would not be every *woman* for herself while she was in the Senate.

Mikulski knew that an entire senatorial career could be stymied in the early weeks of a first term if the senator failed to secure a committee assignment that could have a positive impact on her constituency. Even starting out with a poorly organized office could bury good intentions in chaos. Mikulski drew from her own experience in the Senate, and also from the principles of community organizing that she had learned at home in Maryland.

The most important first step, Mikulski told her colleagues, was to clearly articulate their organizing principles. "The campaign themes that brought you here, such as 'a proven leader' or 'a voice for change,' are not organizing principles," she explained. "They won't help you decide which staff positions you need to fill first, or how to respond to constituent mail. Organizing principles define your overall goals or objectives as a senator. They should reflect your values, concerns, and strengths. A clear set of principles will provide you and your staff with a yardstick. It will help you measure how issues and the decisions you make about them serve your goals, or hinder them."

She showed them her own secret weapon—a three-by-five card that she and her staff always keep with them, and refer to frequently.

BAM'S PRINCIPLES

1. I am not only the senator from Maryland, but also the senator *for* Maryland.
2. We must be committed to looking out for the day-to-day needs of Marylanders and the long-range needs of America.
3. My economic purpose is to help those who are middle class stay there, and to give those who are not middle class the chance to get there.
4. Our constituents have a right to know, to be heard, and to be represented.
5. Listen to the people and the stories of their lives. My best ideas come from the people.
6. We must Communicate, Coordinate, and Cooperate.
7. We are not a bureaucracy.
8. We cannot always guarantee an outcome, but we can guarantee an effort.

9. Always be clear about: "What is the objective we seek?"
10. Goals should be specific, immediate, and realistic.
11. Just move it.
12. Do not explain an abstraction with an abstraction.
13. The language of Washington is a foreign language. We need to talk about people in terms they understand.

The four new senators were thrilled to have her advice, and soaked up all of the information that Barbara Mikulski offered. "I couldn't believe it when I found out there was no formal training provided on how to perform the work of a senator," Patty Murray says. "Even though the rules and procedures are incredibly arcane and complicated, we were just thrown into the mix. Barbara Mikulski was a godsend. She brought all of us together and laid it all out. She explained, 'Here's how you can be successful.' She explained everything—how to get an incorporation into a bill, what you had to do to steer yourself onto a good committee, how to set up a mailroom. Barbara didn't want to be the only Democratic woman in the Senate. She's glad we're here. And we don't want to be the only women in the Senate, either, Democrat or Republican. We want to make sure others get here, too. And the best way we can ensure that it will happen is for all of us to be successful."

It wasn't only Democratic women who benefited from Mikulski's counsel. In 1996, Republicans took control of Congress, and it was a tense period. Fearing a loss of civility, Mikulski invited Kay Bailey Hutchison to lunch, saying, "Civility must start with us." During the years that followed, as each new woman sen-

ator arrived, Democrat or Republican, she was also the benefi-
ciary of the Mikulski welcoming seminar.

There was never any thought to forming a women's voting
bloc. There was a clear understanding that each senator was independent. The fact that they were women certainly didn't mean
that they held the same views on issues—even those seen as traditional women's issues. Kay Bailey Hutchison and Barbara
Mikulski were often light-years away from each other in terms of
political philosophies and agendas. Mikulski's supportive effort
was made without the intention of creating a "women's side" to
the issues that came before the Senate. Her vision extended beyond specific issues or individual pieces of legislation.

Her point was that it was important for women to be heard at
the highest levels of government. Fifty-two percent of the population depended upon the Senate women to give voice to their
concerns—the citizens who also happened to be mothers,
daughters, wives, grandmothers, and the primary caretakers of
elderly parents. In style and philosophy, the new senators were as
diverse as America itself, yet the experiences that formed their
ideals were often quite similar. The practical daily problems that
confronted women—issues related to careers, wages, housing,
day care, health care, education, and public safety—were of
burning urgency to them. These issues had often been pushed to
the back burner in male-dominated senates and legislatures, especially when they involved matters that were gender specific—
such as job equity, sexual harassment, maternity leave, and rape
shield laws.

When Olympia Snowe came to the Senate, she brought a
wealth of experience, with her sixteen years in the House and ten

as co-chair of the Congressional Caucus on Women's Issues, where she broke ground on issues ranging from breast cancer to osteoporosis to Alzheimer's disease and fought for greater equity in funding and in representation in clinical trials.

In 1990, Snowe, along with Democratic representatives Pat Schroeder and Henry Waxman, had asked the U.S. General Accounting Office to look at the exclusion of women from clinical research. "The results were startling," Snowe says. "Women were being systematically excluded from clinical study trials. One study, examining aspirin's ability to prevent heart attacks, examined 20,000 medical doctors, and not one of them was a woman. Even more mind-boggling was a now-infamous study on breast cancer that examined hundreds of *men!*"

OLYMPIA SNOWE

There was a time in America when society accepted deadbeat dads, and child support was considered a woman's problem; when a husband could cancel his wife's pension without her knowledge; when economic equality pertained only to equality among men; when women's health was a page missing from America's medical textbooks and women were excluded from clinical medical trials.

We set out to change this, and we did. We developed the Women's Health Equity Act. I wrote legislation that created an Office of Women's Health Research at the National Institutes of Health. We accomplished tremendous advances for women. We did not allow our differing views on abortion or our partisan af-filiations to get in the way. We didn't agree on everything, but we shared a common vision. We made a difference, there is no doubt about it.

The women of the Senate are a diverse group, but when they come together over a gender-related issue, their collective voice resonates. One of the most striking examples occurred in 1997, when a National Cancer Institute panel charged with providing guidelines for routine mammographies suggested a change in policy. A controversy had developed among health agencies as to when women should start receiving regular mammographic screening. The National Cancer Institute had always recom-mended that between the ages of forty and forty-nine, women should be screened every one to two years, and after fifty, they should be screened annually. However, the findings of a large Canadian study showing that mammographic screening of women in their forties did not lower the death rate from breast cancer led many to conclude that mammograms before the age of fifty were unwarranted. Although many experts thought the study was flawed, it made an impact. There was fear that it would have a direct impact on health-care funding and insurance al-lowances. When an NCI panel suggested that the costs out-weighed the benefits of mammographic screening for women in their forties, it appeared that mammograms for women under fifty would not be included as a part of government standards.

Olympia Snowe asked Barbara Mikulski to join her in a bipartisan resolution calling on the NCI to reissue guidelines encouraging screening for women in their forties. All of the women senators took to the Senate floor in support of the resolution. It passed 98–0.

"In 1955, my mother died of breast cancer at age thirty-nine," says Snowe. "At that time, about one in fourteen women developed breast cancer in their lifetimes; today the number is about one in eight. When the NCI panel made its dubious claim about mammograms, I knew we had no choice but to get involved."

During that period, Hutchison and Mikulski would collaborate across the aisle on another crucial piece of legislation of great importance to women—the Homemaker IRA. This was a true example of heartland legislation, designed to meet a felt need.

The Federal Tax Code stipulated that a single-income married couple was limited to a deductible IRA contribution of $2,250 a year—$2,000 for the working spouse, and $250 for the homemaker. However, if both spouses worked outside the home, each was permitted to deduct up to $2,000, for a combined annual contribution of $4,000.

Mikulski and Hutchison thought the regulation seemed patently unfair, and it was in direct conflict with the nation's values. "The belief that work inside the home is every bit as important as work outside the home is something most of us share," says Hutchison. "Yet we were penalizing women—and an increasing number of men—who chose to stay home and raise their children."

Mikulski also felt that the restriction was counterproductive to the goal of ensuring that older women be financially independent. "Not all work is done in the marketplace," she said.

"When we talk about productivity in the workplace, we need to remember that women are raising the next generation of productive workers. Yet we often don't count what *counts*. The Tax Code wasn't a recipe for a relaxing retirement. It was a plan for poverty."

The Hutchison-Mikulski bill was introduced with the signed backing of fifty-seven senators, representing both parties. All of the Senate women signed on. But perhaps most remarkable was the list of organizations that gave their support to the legislation. It was the first time anyone could remember groups with such radically different ideologies being in such complete agreement on a women's issue. Hutchison entered their letters of support into the record: the Christian Coalition and the American Association of University Women; the Eagle Forum and the National Women's Political Caucus. It was a consensus. The Homemaker IRA, which allowed a married couple to deduct the full $4,000, even if one spouse was not earning an income, was signed into law in 1996.

Many of the most pressing issues that face our nation are not gender specific, but they are part of a domestic agenda that is of primary concern to women. As Dianne Feinstein eloquently stated in her address to the Democratic convention the year she first ran for the Senate: "Our opponents deride this wave of change as 'gender politics,' but it's not about gender. It's about an agenda—an agenda of change: an economic plan to grow our economy, to create jobs, to protect our quality of life, and to protect a woman's right to choose."

There would be times when individual women senators came together, across party lines, in support of a common goal that was not gender specific, but which had important implications for families. One notable example took place in 1999.

BARBARA MIKULSKI

The Congress was getting ready to adjourn. We were fighting on the balanced budget Medicare givebacks. I had really gone out on a limb with my constituents. I'd told them, "I'm going to do everything possible to stop the Senate from adjourning until we put money back into the Medicare program—even if it takes a filibuster."

Of course, once I'd said that I would filibuster the motion to adjourn, I figured I'd better find out if I could do that. So I went to the parliamentarian, and he said, "No, you cannot filibuster the motion to adjourn."

Okay. So now what could I do? I knew that Maine had been terribly hard hit by the balanced-budget cuts. They were facing enormous difficulties in home health care, just as we were in western Maryland. So I went to my colleague Susan Collins, and we formed a little coalition across party lines. We decided to launch a petition drive. I don't think there had ever been a petition drive mounted against a motion to adjourn in the history of the Senate. We agreed that Susan would sign up Republicans, and I would sign up Democrats. Within forty-eight hours, we had sixty signatures. We stood on the Senate floor, and we presented the results to the majority leader Trent Lott, and the minority leader Tom Daschle. We said, "If the Senate fails to restore the Medicare cuts, we have sixty senators who have agreed to vote no on a motion to adjourn."

The petition drive was an old community organizing tactic. You show that you have the power to stop an institution or a road or whatever force you have to stop. Susan Collins and I had the power that Lott and Daschle didn't have. We had sixty senators prepared to go to the mat to protect Medicare. And in the end, we didn't have to stop the Senate. The cuts were restored. That's the power of two women building a coalition to accomplish a mutual goal.

Olympia Snowe points out that the success of women in the Senate is often a result of a collaborative spirit, which comes more naturally to them.

OLYMPIA SNOWE

We certainly don't want to communicate that all of the women in the Senate are homogeneous, with the women sitting on one side and the men sitting on the other. That would be a pretty antiquated point of view, and that's not the case at all. Every one of us is different—in our political positions, our styles, our life experiences. However, women just come from a different place than men do in terms of being more relationship-oriented and more collaborative. In fact, many of the skills women develop in life actually work pretty well in this institution. When you think about it, maybe women are particularly well suited to the Senate, where collaboration is an essential ingredient in getting things done.

When Snowe first arrived in the Senate, she realized that of her ninety-nine Senate colleagues, thirty-four had formerly been members of the House of Representatives during her tenure there. She had already established good working relationships with many of them. "The Senate is an institution where friendships and relationships can make the difference between success and failure on legislation," she says. "Clearly, the fact that so many of my Senate colleagues had previously served with me in the House helped me to hit the ground running in the Senate. Over time, those relationships have helped achieve sound bipartisan legislative accomplishments."

Collaboration and relationships have always been the underpinning of every successful negotiation and resolution arrived at in the Senate. Women come to the task having been conditioned to walk softly but wield their power just the same. It was a lesson Blanche Lincoln learned from childhood—a bedrock Southern value that has served her well.

BLANCHE LINCOLN

Mother used to always say, "If you want a man to treat you like a lady, you need to treat him like a gentleman." If you want your colleagues to treat you like a colleague, you treat them like colleagues. You try to build a friendship and a working relationship. You demand of them the respect that you want by showing up at the meetings, being prepared. When you approach your colleagues from that standpoint, they're usually very appreciative and willing to work with you. And I try very hard to do that. I try to get myself into places where I can be effective for the people I represent. I approach my colleagues and say, "I've got a lot of people near and dear to my heart, and I'm here to represent them, and I would like to work with you on it."

Defining Moments—Washington, D.C., 1993

My gut said, If I want to make a difference for the future, I've got to speak out on this now.

PATTY MURRAY

Patty Murray was a new senator—a freshman, with all the lack of clout the label implies. She was eager to start making a difference, but she didn't want to blow herself out of the water on her first try. That possibility was very much on her mind as she struggled with a decision about what to do—if anything—about a matter of pressing personal concern. In public life, Murray knew, one had to pick one's battles and choose carefully when to put oneself on the line. She also knew that the demands of conscience sometimes battled with the expediency of political survival, and this seemed to be one of those times.

The issue before her now was related to the Tailhook scandal of 1991. Murray vividly recalled how disgusted and angry she had been when the incident at the convention of naval aviators came to light. At the Tailhook Annual Symposium in September 1991, eighty-three women and seven men were assaulted during a rowdy after-hours party. Many women officers testified that their clothes had been torn off of them, and they had been forced to run through a gauntlet of drunken, groping men. The revelations shocked the country—especially when it was learned that the incident was typical of the kinds of activities that had occurred in prior years. By the end of the investigation, 119 navy and 21 marine corps officers had been referred for possible disciplinary action. But after the public outcry died down, nothing was ever done to punish the offenders. To the contrary, many of them were being promoted and were allowed to retire.

The Senate had never challenged any of the offending officers' retirements. No one had ever made an issue of it. But now a former commander of the navy's Atlantic Fleet was retiring at four stars. He had been accused of failing to protect a female lieutenant from reprisals by other officers after she made accusations of sexual harassment. Murray's dilemma? Should she make an issue of it or keep quiet?

PATTY MURRAY

The risk of speaking out was huge. I was a new senator and a woman. I didn't want to get labeled as a flamethrower, and I didn't want to alienate my colleagues before I'd even had a chance to work with them. But my principles were also at stake. I knew that if I wanted to make a difference for the future, I had to speak out.

At a Senate Democratic retreat, I went to the other women senators, and I told them, "I can't keep quiet about this." The others had already reached the same conclusion. Barbara Boxer, who was serving in Congress during the scandal, told about the indignities Pat Schroeder faced when she pushed relentlessly for an investigation. It was crucial that we send a message, as a united group, that we would not sit by and do nothing. So all of the women got together and challenged the retirement of the commander at four stars. We all risked something, and we made a lot of people very uncomfortable. But it was the right thing to do. We didn't win, but we didn't stay silent.

7

Consuming Passions

My single greatest accomplishment, and I mean this quite sincerely, is representing hundreds, thousands of heretofore nameless, faceless, voiceless people. The letters I enjoy most are from those who write and say, "For the first time I feel there is somebody talking for me."

BARBARA JORDAN, FORMER CONGRESSWOMAN

April 2000–Hart Senate Office Building

The door to Senator Dianne Feinstein's private office in Washington, D.C., is usually wide open. She has never been one to stand on ceremony, and her aides move in and out freely as she works at her desk, adding more folders to the already towering stacks that fill every square inch of available space. Actually, sitting at her desk is a rare luxury, and one that Feinstein seldom manages. Like all of the senators, her tightly packed days are typically conducted on the run, her daily schedule broken up into committee hearings, votes, constituent meetings, caucuses, and staff meetings, as well as the charity dinners and ceremonial events that are an integral obligation of public life.

The majority of Americans form their impressions of life in the United States Senate from the flamboyant portrayal of its fights, which play much better on the evening news than the unspectacular pictures of its consensus building. Strife equals ratings, and in the era of the sound bite, that strife is likely to wear a most unflattering label. It's no wonder that so many people in the country accept the idea of the "Do Nothing Congress," whose members can't wait to get out of town for some R&R back in their home states. But if anyone believes that a senator's days are made up of dozing through hearings, making speeches, and attending state dinners, they are mistaken. Each day is long and full—composed of the almost imperceptible steps that lead, with patience and perseverance, to change. Going home—which even the West Coast senators try to do almost every weekend— is hardly a picnic. That is, unless the picnic happens to be a constituent event. Returning to their home states isn't an escape, but a chance to encounter the people they represent, and deal directly with their concerns. Often, senators will say that it is meeting with the people of their state that recharges them. It is a tonic. They couldn't endure the procedural swamp of Washington without keeping in close touch with the people who sent them there.

One of the greatest revelations about Washington is the *local* feeling that is evident everywhere in the offices of the senators. Although situated in grand federal buildings, colorful state flags and emblems deck the hallways. Visitors from home wander into the offices without appointments, and are ushered in for a smile and a handshake with their senator. Staff members set aside their work to conduct spontaneous tours of the Capitol building.

Weekly constituent breakfasts welcome anyone from home who happens to be in town to drop by for a bite to eat and a leisurely conversation with the senator and her staff. Guest books and photo albums are on display, filled with notes and pictures from the folks back home. There is no evidence of the out-of-touch, imperious governing body so often referred to in stories critical of the Senate. Instead, an atmosphere of welcome and openness prevails.

Dianne Feinstein long ago learned to trust her instincts when it came to choosing which political battles to fight, and she is attuned to the issues her constituents care about. When an individual or a group of constituents meets with her in her Washington office, she gives them her full attention, even if it's only for ten minutes. "It's a matter of public trust," she explains. "This may be the only time in their lives that the people I'm meeting with will have a chance to speak directly to someone at this level of the federal government. As a senator, I see myself as the go-between for that government and my constituents. I want them to know that we *do* listen. That doesn't mean we can solve every problem that's brought to our attention. But when they leave here, they go away knowing that they were heard."

Feinstein is a stickler for accountability to her constituents. An example is the carefully monitored system she has set up for responding to their letters. Her office regularly receives up to twenty thousand pieces of mail a week. It is a staggering task to answer each and every one of them in a thorough way—not just with form letters, but with *responses.* That can mean intervention, research, locating records, recommending agencies, giving advice, and explaining the progress and implications of a particular piece of legislation. The correspondence is assigned to a dozen aides, with a goal of preparing at least thirty letters a week each. Feinstein receives a weekly report from all seventy staffers, which details casework, legislative progress, meetings with constituents, and other issues. She reviews these carefully to make sure that people are being helped in a timely manner.

"It's one of the most important things we do," she says. "We respond to constituents. If someone has only answered ten or eleven letters in a week, I want to find out why. The person may have been out sick, or on vacation. But I insist on responding promptly, because it's a matter of accountability to our constituency."

Later, at a staff meeting in her small, packed conference room, she spends a few minutes going over the previous week's numbers, and reiterates the importance of expediting responses to constituent requests. Feinstein runs her office much in the manner of a corporate manager. It's a style that suits her, and for which she offers no apologies. "I'm a creature of local government," she says. "Government has never been an abstraction to me. Whether you're in San Francisco or Washington, D.C., it comes down to the same basic questions: 'What's the problem? What can we do to help?' My father taught me to enforce a certain rigor in my thinking. He was my mentor in that regard. He could reduce the most complex ideas to an essential truth. He'd say, 'Keep probing. Get to the vital truth. Give me the short version: What's the problem?'"

The consuming passion of Dianne Feinstein's tenure in the Senate has been gun control. It is a passion that has been burned into her over many years of personal experiences with the brutal consequences of firearms. Now, in the closing years of the twentieth century and the start of the twenty-first, the issue has crystallized. An ever-mounting catalog of horrors resulting directly from the use of guns has bred a rancorous atmosphere of both contentious debate and increased violence. The rampant availability of guns in our country has always been at the core of the problem. Worse, guns are increasingly being found in the hands of children—witness the series of school shootings that have seared themselves into the national consciousness. "We now have something happening in the United States that has never happened before," Feinstein says grimly. "That is, children killing children with guns. Over thirteen a day." Feinstein has been relentless in the battle, presenting a constant challenge to the NRA—including landmark legislation designed to prohibit the manufacture and possession of military-style assault weapons, the Gun-Free Schools Act, and provisions to ban the import of high-capacity ammunition magazines. She is currently involved in writing legislation that will establish licensing and record of sale requirements for gun ownership. Such legislation, she believes, is the eventual solution to ensure that guns are only in the hands of responsible citizens and out of the hands of children and criminals. Not surprisingly, the National Rifle Association is vehemently opposed to Feinstein's efforts—and she to theirs. Of the NRA's current president and celebrated spokesman, film legend Charlton Heston, she says, "I liked him better as Moses."

DIANNE FEINSTEIN

This is not an esoteric subject for me. My life experience is such that I have seen firsthand what guns have done. I've walked in on robberies. I have seen a proprietor, his wife, and their dog killed by a robber, even though they offered no resistance. I became mayor of San Francisco as a result of assassination. In 1976, a terrorist group called the New World Liberation Front, which had been responsible for blowing up a series of transformer stations in California, placed a bomb in the window box outside my daughter's room. It failed to explode only because the temperature that night was unseasonably cold and the explosive gel didn't function. As a supervisor, I had no protection, so I got a gun permit and learned to shoot at the Police Academy. I carried a .38-caliber pistol for a brief period while my husband was ill and dying. When I became mayor, I succeeded in passing a measure banning handguns in San Francisco, and we instituted a ninety-day grace period for pistol owners to turn in their handguns without incurring penalties. At that time, I turned in my pistol. That pistol and fourteen others were melted down and sculpted into a cross, which I presented to Pope John Paul II during a trip to Rome later that year. The point is, I know where guns work for protection, and I know where they don't. I've lived a life that has been impacted by weapons, so this is not an esoteric, academic exercise for me. Nor is it a political exercise. I come to this issue because of real life experience.

This is a fact: The possession of guns has grown exponentially in the United States. And there's a trickle-down effect to kids. We know it. We've seen it with five- and six-year-old children. For example, the five-year-old boy in Memphis whose kindergarten teacher gave him a time-out because he was acting up, and she made him sit in a corner away from the group. That night, the boy went home, and his grandfather had a loaded gun on the dresser—no trigger lock, no safety lock. The boy put it in his backpack, loaded, and took it to school the next day to kill the teacher who gave him the time-out. That's the trickle-down of guns in our society. I think we've reached an apex of that trickle-down when young kids are bringing loaded guns to school.

Plugging that loophole is critical. That's what I'm trying to do with this legislation. I go all over my state, speaking about this. I ask people, "Does anyone think that a juvenile should be able to buy an assault weapon?" Everybody says, "No." And I ask, "Do you think that the loophole that allows kids like the two youngsters at Columbine to buy those guns, no questions asked, at a gun show should be closed?" Everybody says, "Yes." And I ask, "Does anyone think that a weapon should be sold without a trigger lock?" Everybody says, "No." I have *never* in any speech had a different response. It's just common sense.

Our legislation is also common sense. It sets up a system of licensing and a record of transfer for every firearms owner.

The NRA says that these regulations interfere with basic freedoms. I say they are no more than we require of people who want to drive a car. And how many five-year-olds take the family car and cause death and destruction?

A great myth has been perpetrated by a misreading of the Second Amendment to the Constitution. The amendment reads: "A well regulated Militia being necessary to the security of the State, the right of the people to keep and bear arms shall not be infringed." The interpretation of the gun movement places the entire emphasis on the last part of the sentence. As a result of this deliberate misreading, we have more guns and a higher homicide rate than any other industrialized nation. We have high-powered weapons of war being sold freely to children.

There is no question about the power of the NRA and its supporters. On the other hand, there is no question about the rightness of our quest.

In her weekly staff meeting, Feinstein talks about organizing support for the planned Million Mom March, scheduled for Mother's Day, May 14, 2000, in Washington, D.C. "Let's get behind this," she exhorts her staff. "It's about time women rallied to say enough is enough to the gun manufacturers. Do you realize that many handguns have so little trigger resistance that a three-year-old child can manage to fire them? A three-year-old can pull the trigger on a gun. It shocks the conscience." She holds up a flyer from the NRA. "Now look at this," she marvels. "On the same day as the Million Mom March, they're having an Armed Mom March, right out of Fairfax, Virginia."

The staff meeting consumes one hour from start to finish, and then Feinstein is up and moving again. A group of constituents is waiting in the front office. She comes through the door, smiling warmly, extending her hand. "It's so good to meet you," she says. "Call me Dianne."

Feinstein loves her work, and relishes the platform the Senate provides. If she has a major complaint, it's about the relative disorganization and lack of rules for which the Senate is so well known. She readily acknowledges that it drives her crazy when she can't control her schedule. "Ah," she sighs ruefully, "if I were running the place, things would be different."

Dianne Feinstein was once asked if she liked being in public life. "*Like* it?" She rolled the word around on her tongue as if it were a drink with an unfamiliar taste. She finally answered it this way:

DIANNE FEINSTEIN

In 1981, I had the opportunity to be seated next to Indira Gandhi at a dinner in Manila. I asked her if she enjoyed being prime minister of India. I've never forgotten her answer. It was a careful response, and very honest. She said, "I was born to a time and to a family. Perhaps I would not have chosen it. I had no choice. But it has been very rewarding." So, if you're asking, "Is this what I am meant to do?" the answer is yes. Is it my calling? Yes.

While it's gratifying to be able to have a positive impact on their local communities and help solve the problems facing them, the issues that the senators deal with often produce wide-ranging results that resonate on a national level. They are very conscious of the need to balance the responsibilities of the federal government and the responsibilities of the citizenry. Constituent meetings are not merely opportunities to listen to the needs of supplicants. The meetings are fashioned as problem-solving workshops: *define the issue, describe the impact, articulate the goal, brainstorm strategies for a solution.* Within this context, the question of the federal government's participation is raised in a very specific way. *Does the federal government have a role in this problem? How can we find a way to be partners in the process of resolution?*

The answers are not always so clear-cut. Perhaps one of the most difficult challenges a senator confronts is finding a way to balance the altruistic desire to help people in need with a pragmatic assessment of what is possible. Sometimes a senator is placed in the position of having to deliver disappointing news— especially when a local group has applied for federal funding.

When matters arise that concern public safety and well-being, senators are often compelled to step forward and act as guardians to those who have been affected. In the course of a year, senators conduct a number of investigations, and hold hearings on matters that directly concern their constituents.

March 1999–The Capitol: Senate Hearing Room

As Susan Collins arranged her papers on the desk before her, preparing to chair the hearings that followed a year-long investigation into deceptive sweepstakes mailings and promotions, she paused for a moment, looking out over the crowd taking seats in the hearing room. Lined up in the front rows, looking distraught, were the elderly men and women who would testify today, along with some adult children who would speak for their deceased parents.

Long before she came to the Senate, Collins had established herself as a fierce defender of consumers. During the years she served in Maine governor John McKernan's cabinet, she had been in charge of regulatory agencies and boards and had made consumer protection and restitution her top priority. Seeing that people, particularly senior citizens, are treated fairly matters deeply to Collins. She abhors the scam artists and unethical businesses that prey on the vulnerable and the trusting.

Collins brought this same commitment to the Senate. When it was announced that she would chair the Permanent Subcommittee on Investigations, the Senate's only investigative subcommittee, Collins decided to make its focus consumer protection and advocacy. Under Collins's leadership, the committee has uncovered and investigated Medicare fraud, Internet scams, telephone slamming, and other practices that ripped off or were unfair to consumers.

When Collins first began to study the problem of deceptive sweepstakes practices in 1999, she was initially surprised to learn of the shattered dreams and broken lives that could result from something as seemingly innocuous as a sweepstakes mailing. She has learned a great deal since and is no longer surprised—just resolutely determined to do something about it.

Over the course of two days, Collins and her subcommittee members listened to heartbreaking testimony from the victims of sweepstakes mailings. Some of these individuals had spent thou-

sands of dollars, sometimes going bankrupt, in the hope and ex-
pectation that they were about to win millions of dollars. The
committee heard the sincere, shamed voices of the duped senior
citizens. Some had been so certain they were the chosen winners
that they refused to leave their homes on Super Bowl Sunday,
fearing they might miss the Publishers Clearing House Prize Pa-
trol that came to the winner's door and presented them with a
check for $10 million. The committee listened to testimony
from experts who demonstrated how sweepstakes companies targeted elderly citizens with a flood of provocative mailings and
phone calls. And, finally, they heard from the representatives of
the sweepstakes companies themselves. They defended their mar-
keting approach, of course, and attempted to reassure the committee that the sweepstakes conditions were perfectly clear.

Susan Collins was known for maintaining an objective, even-
handed demeanor during hearings. Above all, she believed it was
important to be fair, do your homework, listen to both sides of
an issue, and ask questions that would bring clarity. But she
found it hard not to be deeply affected by the testimony of the
elderly citizens. These were good people, who had worked hard
all of their lives and raised families. They had saved over the
years, and put aside money in retirement accounts and pensions,
so they could continue to provide for themselves, intending
never to be a burden to their children. Beyond the financial dis-
tress into which all of the victims were thrown, there was a collective feeling of deep humiliation that each expressed.

In particular, Collins would never forget one man's plight. Eustace Hall was a sixty-five-year-old retired medical technologist
who testified that he had spent up to $20,000 buying magazine

subscriptions and making other purchases related to sweepstakes promotions. He was so overcome with emotion as he read his statement that his daughter had to finish reading it for him:

> Over the years, I received many personalized letters from the sweepstakes companies thanking me for being such a good customer, and telling me that my chances of winning were good, or that it would be my time soon. I have a copy of a letter from Dorothy Addeo, Publishers Clearing House contest manager. I would like to read a short portion of the letter. "My boss dropped into my office the other day, sat down, and sighed. 'What's the story with Eustace Hall? I see that name on our Best Customer List, on our Contenders List, on our President's Club Member List. But I don't see him on our Winners List. There must be something we can do to change that. It's not right when someone as nice as Eustace Hall doesn't win.' Then he sighed again, looked at me, and left, and I sat there wondering what to do. I had my mission, Eustace Hall, to make you a winner, and soon."

On the second day of the hearing, Collins had the opportunity to question Debbie Holland, who was representing Publishers Clearing House. She was incensed by the testimony of the victims and wanted an explanation for what appeared to be blatantly deceptive practices. Trying to control her anger, Collins said, "Ms. Holland, I have to tell you that I was absolutely stunned by a statement that you made in your earlier testimony.

You said, 'We believe that our promotions are clear, and that no reasonable person could be misled by them.' Well, Eustace Hall is a reasonable person who testified here before us yesterday, and he was completely misled by your mailings. We have had hundreds of consumers from across the United States contact the subcommittee with concerns about sweepstakes, mainly yours and the other companies who are represented here today. They appear to be reasonable people."

Collins held up the letter from Dorothy Addeo to Eustace Hall. "You are familiar with this letter, Ms. Holland. This is a mailing that Mr. Hall brought to the attention of the subcommittee." In a firm, clear voice, Collins read the text of the letter, then lifted her eyes and fixed them on Debbie Holland. "Mr. Hall told our investigators that he believed that this exact conversation took place; that, in fact, Dorothy Addeo's boss did drop by her office and say these words. Did this conversation actually take place with regard to Mr. Hall?"

"It was a dramatization of conversations that did take place when we were planning this promotion. The mailing was sent to nine million people," Debbie Holland replied.

"Did you have nine million conversations?" Collins asked dryly.

"Of course not," said Ms. Holland.

Collins sighed. "But do you not see why this would be deceptive to a reasonable person like Mr. Hall? Do you not see why it made him think that he was special?"

Debbie Holland replied evenly, "We do not think that this is deceiving. We thought it was perfectly fine."

Collins and her colleagues disagreed. On November 19, 1999, the Senate unanimously approved Collins's legislation, the Deceptive Mail Prevention and Enforcement Act, designed to limit the deceptive techniques and claims used by sweepstakes companies. On December 12, 1999, President Clinton signed the bill into law.

SUSAN COLLINS

I'm proud of that bill. That's why I'm here—to help people with problems that affect their daily lives. You might think, "Well, sweepstakes scams, how big an issue is that?" Let me tell you, it matters to families across America who have lost their money and pride because of deceptive mailings. It's also an example of the impact ordinary people can have on government. The way that I became interested in deceptive sweepstakes mailings in the first place was through constituents who sent along copies of mailings they'd received. They wondered how these companies could announce, in big, bold letters, "You're a winner," when in reality, the person receiving it hadn't won anything at all. And then, so many people wrote in. They were saying that their fathers, mothers, or grandparents would get hooked in, and end up spending thousands of dollars every year on these sweepstakes offers. It was really heartbreaking. It wasn't right for them to be exploited. Those letters led to the hearings and to the legislation. Protection from those deceptive practices is now law. This is the result of people from home letting me know about this problem. That's making your voice heard, isn't it? People shouldn't underestimate the power of their voice!

When she speaks to groups of young women, Dianne Feinstein always urges them to "find your passion. Find that one thing you're good at, that you can dedicate your life to." Patty Murray found her passion early in life; it continues to consume her.

1999—Seattle, Washington

"How many of you are planning on becoming teachers?" Senator Patty Murray asked a roomful of sixty local high school seniors. Not one hand rose in response to her question.

Murray was not surprised. The role of teacher had become devalued of late—just when there was a desperate need for young people to enter the profession. She remembered when she was growing up. The response to such a question then would have been a flurry of eager young hands flung skyward. Back then teachers were still admired. They were to be emulated. They had authority and prestige.

She looked out at the young, earnest faces. No future teachers there? Not a one? She decided to ask again, only this time she posed the question differently: "How many of you would be interested in pursuing a line of work that was held in high esteem, where you were respected by everyone in the community, where you were very well paid, had excellent benefits, and an opportunity to advance—a line of work that offered creative and intellectual satisfaction and gave you a chance to make a tremendous difference?"

Now every hand in the room was raised. Interest was suddenly high.

"Good. Now put your hands down for a moment. Sounds like a good job, doesn't it?" Murray smiled, leaning forward on the podium. "If the line of work I just described to you happened to be teaching, how many of you would still want to do it?" At least half the seniors raised their hands again.

"That's wonderful. Thank you. I hope all of you do go on to teach someday. Now imagine how strong our nation's schools would be," Murray said, "if half of all high school seniors aspired to someday be teachers."

Murray's primary passion has been public education. She believes that a strong educational system is at the core of virtually every other issue in government. "It has been our commitment to public education that has strengthened the core principles of our nation's democracy and has spearheaded all of the enormous progress we've enjoyed over the last century," she explains. "If we expect to bring a new generation forward into the next century, then we have an obligation to provide them with exceptional educational opportunities. And that means strong public schools, more teachers, and smaller classroom sizes."

In the Senate, Murray had been known for her commonsense approach to improving public education. Given her experience in the classroom, she knew firsthand that overcrowding made it more difficult for students to learn. So in January 1998, when President Clinton proposed the idea of hiring 100,000 new teachers and paying for them with a new tax on tobacco, Murray was galvanized. This was legislation she was willing to go to the mat for. If successful, it would support one of her most important education goals: the reduction of class sizes.

The road to legislative success is circuitous, to say the least. A bill rarely gets passed without undergoing a series of defeats and revisions. Few citizens can fully appreciate the level of tenacity required to push a bill through Congress. When the legislation in question calls for putting money into the coffer to hire teachers, the task is truly daunting.

Murray's legislation was brought before the Senate twice. It was defeated twice along partisan lines. In the meantime the Clinton proposal of a further tax on tobacco was defeated as well, robbing Murray's bill of its original source of direct funding.

Even with all those negatives mounting against her, Murray wasn't done yet. She had one remaining angle—finding seed money in the federal budget. Finally her efforts paid off. Murray was able to gain a one-year appropriation of $1.2 billion to hire and train twenty-nine thousand new elementary-school teachers for kindergarten through third grade across the country. It was an important victory for Murray, but it was only the beginning. To truly achieve a measurable success, Congress would have to continue its commitment for more than just a year. Murray was looking for long-term, systemic change, not just a one-shot measure that allowed the hiring of a few thousand new teachers.

When she sought authorization in September 1999 for a further $1.4 billion in the labor, health and human services and education appropriations bill, her amendment was tabled.

This was precisely the kind of legislative behavior that so many of Murray's constituents complained about, and rightly so. How could progress be sustained without a commitment for long-term change? So Murray pulled out her most reliable strategy. She started working the phones. By October, she had the backing of thirty-seven fellow senators. They sent a letter to President Clinton urging him to veto the appropriations bill if it did not fully fund the class-size reduction initiative. He did so, knowing that the caucus Murray had skillfully put together would sustain his veto.

In mid-November 1999, during the final hours of budget negotiations between the White House and Congress, Murray's initiative made it back into the package. Her amendment was allotted $1.325 billion. It was a much-sought-after victory, but Murray was unable to savor it fully. She didn't have the time. Education reform was going to be a brick-by-brick building project.

Every issue has a human face. It ultimately reduces itself to the way it impacts on a real person, though it may be surrounded by a host of complexities. Perhaps the greatest gift women in the Senate have brought to the process has been their ability to personalize constituent concerns—to see the human, very vulnerable face beneath the imposing facade of law and politics. There are big problems and seemingly small problems, but all are the same. Each matter brought before the Senate is scrutinized through the focused lens of representative government.

March 1999—The Capitol

For Senator Blanche Lincoln, the issue was exquisitely simple: Thousands of farmers in rural Arkansas had purchased crop insurance from a company that was going to renege on its contracts and leave them high and dry. With rice prices expected to plummet for the coming year, farmers were concerned that they'd need additional help, and American Agrisurance, Inc., was promising a return of three cents per bushel above what the federal government usually offered. Farmer interest and participation in purchasing the coverage was larger than the company expected and a larger financial burden than it could bear.

American Agrisurance (AmAg) realized it was going to lose money and on March 1, 1999, told its agents that it would no longer honor a three-cent return and could offer only one and a half cents per bushel. On March 10 it instructed agents to halt all sales of the policy.

Unfortunately for Arkansas's farmers, the deadline for purchasing the federal government's crop insurance program had already passed. Thousands of farmers who had relied on policies from AmAg were now in danger of being left without adequate insurance. Farmers went ballistic; they knew rice prices were going to be low and they needed as much assistance as possible to stay in business.

Blanche Lincoln had never been accused of being a shrinking violet. She preferred compromise to conflict, but she could fight hard when circumstances demanded it. Now she had to take a strong stand on an issue of pressing concern to her constituents. Lincoln, the daughter of a seventh-generation farm family, took their concerns to heart and immediately went to bat for her state's agricultural producers.

During a series of congressional hearings, she expressed displeasure and frustration on behalf of her constituents. At one Senate hearing she revealed an internal American Agrisurance memo, obtained by her staff, in which company executives touted their success in selling the policies and urged employees to "look forward to Maui to celebrate your hard work this year." Lincoln was outraged. "Does this sound like a company in dire straits to you?" she asked colleagues.

Lincoln also contacted the secretary of agriculture, Dan Glickman, urging him to intervene and to reopen the government's window for farmers to purchase federal insurance. She introduced legislation to open up the policy cancellation period so rice producers who weren't comfortable with AmAg's decision could choose another method of coverage. Lincoln sent letters to her colleagues urging their support and made her pitch personally to senators in committee and on the Senate floor.

"At a time when the agricultural climate in Arkansas is devastated to begin with, these policies were a last ray of hope for hundreds of farm families," Lincoln said on the Senate floor. "This company pulled the rug out from under our farmers, and I understand that in a situation like this, it's the American farmer who gets stuck holding the bag. Every year farmers go out on a limb and make critical planting decisions based on obligations and promises. Rice producers in Arkansas bought these policies in good faith. They deserve every opportunity to make the best financial decision for their farm and get back on their feet."

Lincoln's hard work paid off. The leadership from both parties and the top members of the Senate Agriculture Committee agreed to support her bill. Both the Senate and the House of Representatives unanimously approved the measure. The Department of Agriculture agreed to extend the deadline for farmers to purchase federal crop insurance coverage. And President Clinton signed Lincoln's bill into law on April 1, 1999—just thirty days after the company announced its change in coverage.

BLANCHE LINCOLN

I realize that many people feel cynical about government. That makes it especially gratifying to me when we can actually find practical solutions for their problems. When we're responsive to people's needs, we begin to build a sense of pride—give people a reason to look up to their government, as opposed to grumbling about it.

I had that experience recently. I got a letter from a guy back home—someone I'd heard complaining many times. You might say he was cynical about government. He wrote, "There's an old saying, 'I'm government and I'm here to help.'" And he told me the saying was used in a derogatory way—sarcastically. He said that he had often used it that way himself. But this year, our office was involved in trying to solve a matter in his area, and we reached a conclusion. He wrote that he'd never think of the saying, "I'm government and I'm here to help" in a derogatory way again—and he'll challenge anyone who does. Well, that letter made my day. Every once in a while, you can really change someone's attitude and show them we mean what we say when we talk about constituent service. Providing help on these practical problems is such an important part of the job. I know that the local issues aren't glamorous, so they don't get much media attention, but, believe me, that's where the reward is. You make sure that hundreds of farmers can stay in business—that's results.

Here's another example—child car seats. Now, I believe I have some credibility on that issue because of my boys. So I was glad to cosponsor Senator Fitzgerald's legislation to improve the standards for testing car seats. I was shocked to learn that they're still using testing methods from the 1970s. As a mother, how could I not do whatever I could to help protect the nation's little ones?

Blanche Lincoln considers it her primary mission to educate her colleagues in the Senate—especially those from states that are largely urban and suburban—about the special difficulties faced in rural communities. She emphasizes the fact that for rural communities and families, many of the provisions in current legislation are meaningless. "It doesn't make any sense to put in school voucher programs when you can't get the teachers to go out into those areas," she says. "It doesn't make sense to give disabled persons access to physical therapy when they can't get to the sites. Or if you cut home nursing programs, you may be eliminating the only health care people have access to. Often the provisions are generic, and they're based on a model of access. I think it's important that we give states the authority to use the resources as they see fit, backed by some form of accountability. But don't just pretend that everyone lives five minutes from a town or city."

Lincoln believes that technology will prove to have a transforming effect on rural communities. "Just recently I met a woman who had been a welfare mother," Lincoln says. "She was mildly physically handicapped, and that made it impossible for her to get to a job. Now she's working for a Web site, and she's making a good salary. Think of what Internet-based companies can make possible in rural areas."

It's a passion that is shared by her colleague Barbara Mikulski, who is exhilarated by the technological advances that she believes will be the hope and salvation of so many struggling communities.

December 1999—Baltimore, Maryland

"Stick with me. You'll have a future," Barbara Mikulski states with assurance as she shepherds a delegation from NASA through the doors of a large building near the Baltimore waterfront. This was once the home of the American Can Company, one of Baltimore's largest employers. Today the building houses an information technology training center, which is part of a joint venture between NASA and the Greater Baltimore Alliance.

The renovation of the old American Can Company building is a striking architectural achievement. It has employed the inner workings of the factory—a vast interior space lit by enormous multipaned windows, vaulted ceilings, and giant pipes and tubing now fully exposed—with clear, wide open work spaces to bring about a transformation to a new era of usefulness, a harbinger of the techno-industrial working environment of the future.

The project, called Space Hope, is a key component in Mikulski's new passion: digital empowerment. Space Hope is an innovative workforce training program devised in response to a real need. Workers needed new technological skills in order to avoid being left behind. NASA's Goddard Space Center in Maryland faced a shortage of workers skilled in information technology.

When Mikulski was a child many years ago, the American

Can Company represented the American Dream to countless immigrant families. Today it represents a new kind of dream—one that could not have been envisioned in the working-class neighborhoods of Baltimore forty years ago.

As the ranking Democrat on the Appropriations Subcommittee on Veteran Affairs, Housing and Urban Development, and Independent Agencies, which oversees funding for NASA, Mikulski had the clout to broker this deal. "We're wired and inspired," she tells the NASA reps. "This is a part of the shift to the second industrial revolution—which will be knowledge- and technology-based."

BARBARA MIKULSKI

There is a digital divide in America. Those who have access to technology and know how to use it will be ready for the new digital economy. Those who don't will be left out and left behind. I see digital empowerment as my legacy issue because it will set the tone for the next century. We have the ability right now to make sure that every child in America has access to the tools that will empower them to succeed in the new economy. We can avoid the social chasm that came out of the first industrial revolution, which set up a system of haves and have-nots.

In March 2000, Mikulski introduced the digital empowerment bill, her legislation to close the digital divide, on the floor of the Senate. Her impassioned plea was trademark Mikulski—a fine blend of the strategist and the missionary. Her closing was a clarion call: "I give all praise and thanks to the Dear Lord, who has inspired me to do this and gives me the opportunity to serve in the Senate. I truly believe one person can make a difference. I am trying to do that with this legislation. If we can work together, I know we will be able to bring about change—change for our children and change for the better."

Despite their diverse agendas, one common theme that is repeatedly demonstrated by the women of the Senate is tenacity—a dogged attention to detail. Reflecting on that urge to do something, Olympia Snowe smiles in fond recollection of her upbringing and its work ethic. "My aunt used to say to me, 'Olympia, don't ever wait for me to ask you. Look around and see if there's something that needs to be done, and do it.' So guess what? Now I can't sit still. As soon as I sit down, I find myself thinking, 'What needs to be done?' And then I start doing it. I can't help myself."

8

Out on a Limb

I'm making a speech and you're not going to like it.

It's not as if Margaret Chase Smith was prone to controversy. The gentlelady from Maine was the very essence of a moderate Republican. But as often seems to be the case with moderate Maine Republicans, she was an independent thinker, and she would not go along with activities that she believed were wrong.

In the stories of the Senate's ultimate condemnation of Wisconsin senator Joseph McCarthy for his witch-hunt against suspected Communists, Margaret Chase Smith's name doesn't come up as often as Joseph Welch's. Most of us are familiar with the moment in 1954 when Joseph Welch, the counsel for the army, pressed by McCarthy to betray an officer, wearily uttered the words that finally toppled McCarthy's reign: "Have you no sense of decency, sir, at long last? Have you no sense of decency?"

What is less often mentioned is the speech that Senator Margaret Chase Smith gave in June 1950 in front of the Senate—a full four years earlier. Smith was a freshman senator, barely a year and a half in office, when she made her speech, calling it "A Declaration of Conscience." In it, she attacked McCarthy, without referring to him by name, for using congressional immunity to make unproved charges that defamed innocent Americans. She accused him of dividing the nation with his tactics.

She concluded with a riveting image: "I do not want to see the party ride to political victory on the Four Horsemen of Calumny—fear, ignorance, bigotry, and smear."

Although the speech attracted favorable nationwide attention and was endorsed by six fellow Republicans in the Senate, it did little to restrain McCarthy. His influence would continue another four years. In 1954, when Smith was up for reelection, McCarthy put his support behind Robert L. Jones, her opponent in the primary. Smith defeated him with 82 percent of the vote. In December of that year, she voted along with her colleagues to censure McCarthy.

Smith's courage in 1950 made her the first Republican to condemn McCarthy, an action that most public officials believed was tantamount to committing political suicide. She went out on a limb as an act of conscience, backed by a strong conviction about what government was meant to do, and what it was not meant to do.

Olympia Snowe believes that her predecessor's act of courage, supported by thirty-two years of conscientious service to the state of Maine and the nation, is a fitting model for what every good public servant should be. She has often invoked Senator

Smith's memory when she needed an example of bravery and tenacity. In 1999, Snowe lobbied for and secured the commission of a portrait of Smith to hang on the Senate side of the Capitol. Only two other women have been given that honor—Hattie Caraway, the second woman senator, and Pocahontas.

For today's public servants, going out on a limb often means standing at the center and taking the moderate position. Many elected officials, such as Snowe, have reached the conclusion that intense partisanship and ideology have created a divisive atmosphere in Congress that has made governing increasingly difficult. Moderate senators have shouldered the task of bringing both sides together to reach a point where compromise is possible and results can be achieved.

Snowe would like to see the sensible, results-oriented work ethic returned to government. She originally entered politics because she believed it was a high calling, a chance for accomplishment, and her early days in the Maine Legislature fulfilled that promise.

OLYMPIA SNOWE

In those days, once the election was over, we put the campaigns and party labels behind us so that we could get down to work. Too often these days the campaigning never ends and the governing never begins.

The political dynamic confronting our system is the ongoing erosion of bipartisanship, civility, and cooperation. In many instances, political leaders have failed to seek compromise, and instead approach politics as an all-or-nothing proposition, where there are only two outcomes—a scorched-earth victory for one side, or political stagnation. Such constant friction deadens civic impulse.

In her years in the House and now in the Senate, Snowe has earned a reputation for her principled moderate approach to critical issues. During the bitter 1995 partisan debate over welfare reform, Snowe and others described by *Congressional Quarterly* as "passionate moderates" sought ways to reach consensus. They agreed on the importance of overhauling the welfare system. At the same time, they wanted to prevent a "race to the bottom" that could follow any efforts to remove federal mandates.

OLYMPIA SNOWE

When the Senate began debating the welfare reform bill in September 1995, it was believed by many to be an exercise in futility. But—in small groups—members from both parties set aside their party labels and focused on the first major overhaul since the Great Society.

I worked with Senators Orrin Hatch and Christopher Dodd and then with Majority Leader Bob Dole and several other moderate Republicans to reach agreement on child care, triggering a compromise that allowed the Senate to pass welfare reform by an astonishing 87 to 12 vote. Ultimately, this bill became the foundation for the final reform package signed into law by President Clinton.

It was a demonstration that even when passions ran high, the system could work. And in the five years since Congress reformed welfare, its replacement has helped usher the neediest Americans into futures of greater hope and opportunity.

In 1996, when Susan Collins joined Snowe in the Senate, she represented a second independent voice from Maine. Only six months into her tenure, Collins took a major stand on campaign finance reform. It was an issue that was near and dear to her heart. After running a tough campaign against two opponents with massive war chests, Collins pledged to her constituents that she would make campaign finance reform a priority, and she has kept that pledge, although the reforms have yet to pass.

Collins believes that being a moderate in a field of conservative Republicans can be an advantage. "What I anticipated and found," she says, "is that moderates play a key role in the Senate. And if you are a moderate Republican you have disproportionate clout and influence early on, because it's the moderates in the Senate, both Democrats and Republicans—but particularly the Republican moderates—who hold the balance on most votes. This is particularly true when the Senate is so closely divided between Republicans and Democrats. If a vote is close, the moderates control the outcome."

Collins, Snowe, and other like-minded moderates from both sides of the aisle, pushed hard for the McCain-Feingold campaign finance reform bill, but when it reached the Senate in early 1998, it failed to pass. But the work continues.

In September 1997, Susan Collins led the fight, along with Senator Dick Durbin (D-IL), to repeal a $50 billion tax giveaway for tobacco companies that was secretly slipped into a tax bill at the last minute. It was not in the House or Senate versions of the bill, and there was never any public debate on it. This one-sentence provision just magically appeared at the end of the 327-page bill, tucked into a section entitled "Technical Amendments Related to Small Business Job Protection and Other Legislation." To this day, no one has claimed parentage.

In her remarks on the Senate floor, Collins noted that this was "a stereotypical example of backroom politics at its worst" and that it represented "the kind of abuse of the legislative process that the American people are rightfully sick and tired of, a secret agreement, negotiated behind closed doors by powerful tobacco industry lobbyists in the closing hours of consideration of a massive tax bill."

After the tax break was exposed by Senators Collins and Durbin, the Senate voted overwhelmingly in favor of the Durbin-Collins amendment to repeal the $50 billion giveaway to big tobacco, the House concurred, and President Clinton signed the repeal into law as part of the fiscal year 1998 labor, health and human services appropriations bill.

People in Maine respect the independent voice. According to the Center for the American Woman in Politics, in 1996, of the six Senate races where men and women made different choices, in every case the women favored the Democratic candidate. However, the Republican who came closest to winning the women's vote was Susan Collins.

The Republican leadership has credited Collins with helping to shape the party's direction. In January 2000, she was chosen, along with Senator Bill Frist of Tennessee, to give the Republican Response to President Clinton's State of the Union Address. The visible platform was a demonstration of her party's respect and confidence.

Speaking to a prime-time television audience, Collins detailed the Republican priorities for the coming year.

> Our Republican agenda is driven by the simple but powerful truth that America will continue to lead the world as long as our government allows opportunity, initiative, and freedom to flourish. Letting people create what they can dream has transformed our economy.
>
> As we reflect on our economic health, we should never forget that America's recent success is, above all, a triumph of values. Americans will never let our country become rich in things and poor in spirit. . . .
>
> A good education is the ladder of opportunity. It turns dreams into reality. That's why education is at the top of the Republican agenda. . . .
>
> The debate in Washington is not about money. It is about who makes the decisions. We need a change of approach—one that recognizes that local schools, not Washington offices, are the heart and home of

education. We will empower states and communities to use federal education dollars in the ways children need most. . . .

Education today is America's broadband to the future—a powerful conduit for achievement and success. Let us work together to ensure that all Americans have the educational opportunity for a bright future.

As JFK once said, "The best time to fix the roof is when the sun is shining." We're ready to get our toolbox out and go to work.

MARY LANDRIEU

Their nickname is the Mod Squad, but that's where any similarity to the sixties TV series ends. In this instance, "Mod" means moderate, and the goal of the Senate New Democratic Coalition is to demolish legislative deadlock. The Mod Squad's leader, Senator Mary Landrieu of Louisiana, is a no-nonsense New Democrat, who has never fit very comfortably into any extreme partisan niche. She grows impatient with the Senate squabbles between the far right and the far left that too often lead to stalemate and inaction. Landrieu believes that the Mod Squad's influence can be most meaningful before any legislation reaches the floor. "This isn't just about casting votes. It's about shaping what comes before the Senate," Landrieu explains. "Our goal is to convince our colleagues to write legislation in ways that won't automatically set off alarms on the left or the right."

Joining Landrieu in forming the coalition is Blanche Lincoln of Arkansas. The two women share similar views on how best to govern. Their roots are in states that have always been predominantly Democratic, but tend to be fiscally conservative. Landrieu formed her centrist philosophy as state treasurer; she avoided the flamboyant wrangling of Louisiana-style politics, and formed coalitions among those who shared her moderate views.

In Congress, Lincoln helped create the Blue Dog Coalition, a group of fiscally conservative Democrats. There are several theories as to the origin of the name "Blue Dog." Lincoln mentions one possibility: "A Yellow Dog Democrat is a die-hard Democrat. A Blue Dog Democrat is a Yellow Dog who's being squeezed from both sides." In other words, the one in the middle.

For Lincoln, the shift to blue was an easy decision to make.

BLANCHE LINCOLN

I came to Washington with the idea that I wanted to make changes, and bring new energy and new light to the process. It stems from the years I spent in Washington. Government wasn't results-oriented. It was too reactionary. They'd watch the problems grow and grow, and finally react when it was almost too late. The Blue Dog Democrats were fiscally conservative, and we looked at issues from a nonpartisan perspective. We wanted to get results. We met once a week, and our budget in 1995 became the basis for the balanced budget of 1997—except for the health-care provisions. We believed in putting the facts together line by line on the budget, and were instrumental in welfare reform.

Groups like the Blue Dog Democrats and the Mod Squad indicate to our constituents that we're serious about working on the issues. They also communicate to the leadership that we're independent. We have one goal—to move that ball down the field.

Like Olympia Snowe and Susan Collins, Lincoln, too, has a historic role model. Hattie Caraway of Arkansas, a Democrat, was the second woman to serve in the Senate. She began her Senate career in the traditional manner of women of her day—she was appointed in 1931 to fill the seat of her deceased husband—but then she astounded everyone by making it more than an interim position. When she announced that she would seek reelection to a full six-year term, the party leaders were nearly apoplectic. She won the election after receiving surprise support from the powerful Senator Huey Long of Louisiana, who declared, "We've got to pull a lot of potbellied politicians off a little woman's neck." (Incidentally, Long's own wife, Rose McConnell Long, would be elected only four years later to finish Long's term when he died. She would serve less than a year.)

Hattie Caraway was a quiet but diligent senator, and a strong supporter of labor unions, veterans, and women. She ran for a second full term in 1938, and defeated her opponent, whose campaign slogan was "Arkansas needs another man in the Senate." Caraway went on to become the first member of Congress to endorse the Equal Rights Amendment to the Constitution.

Blanche Lincoln is proud of Caraway's contribution—and her spunk. She identifies with her predecessor's determination that she would not be left outside the door while the business of the nation was transpiring inside.

Landrieu is a pragmatic politician who votes across the spectrum. On various issues she has sided with conservative Republicans and liberal Democrats, but it frustrates her that the choice is often posed as being between two extremes. They formed the coalition after a particularly contentious year in the Senate.

"This group is part of a growing number of Democrats who believe that governing from the center out is the best way to get the job done," she said. "We believe in a balanced budget, debt reduction, accountability in education, and widening the circle of opportunity for more Americans."

MARY LANDRIEU

In 1999, nearly every major debate, from tax cuts to the Patients Bill of Rights, broke down into partisan bickering. What was the result of that? Needed reforms didn't take place. Since the loudest voices in the Senate are usually those on the far left or the far right, I decided that the sensible middle should have a loud voice, too. This is not an assault on the Democratic Party. We're trying to strengthen the party. We're proposing a new approach for our party, but let me emphasize that it is an approach grounded in the traditional Democratic values of opportunity, responsibility, and community. The Mod Squad aims to take a nontraditional approach to traditional values.

But it's not just the moderates who are building coalitions and creating sensible approaches. The women in the Senate today represent diverse political philosophies—they range from liberal to moderate to conservative. Yet while their philosophies on many issues may differ, they share an interest in looking at problems from the standpoint of what Barbara Mikulski calls "felt need." They are less inclined toward political power plays, more inclined toward resolution.

Mikulski notes the close relationship she shares with Senator Kit Bond of Missouri, her Republican counterpart and chairman of the powerful appropriations subcommittee on veterans affairs, housing and urban development, and independent agencies. This subcommittee funds the nation's space and science programs, channeling billions of dollars into high-tech labs and universities across the country, creating thousands of jobs. Mikulski and Bond differ philosophically on many issues, but they share a deep commitment to ensuring that the nation maintains a prominent role in science and technology. For example, in 1999, amid the intense budget battle between Republicans and Democrats on Capitol Hill, Mikulski and Bond were able to resolve their differences and quickly reach agreement on the appropriations bill.

Dianne Feinstein, whose position on gun control places her on one side of the most polarized debate in America today, believes that no matter where you stand on the political spectrum, you can't resolve problems without negotiation.

DIANNE FEINSTEIN

I've been a part of a city that became very polarized. The assassinations in San Francisco took place as a result of that polarization. It was a huge learning experience for me. I learned that when you live in a diverse society, you run terrible risks of polarization—because the smallest thing can disrupt everything. I determined a long time ago that my kind of politics was to concentrate on bringing people together—working out practical solutions. I wasn't going to be the ideologue who didn't enable people to solve problems.

An example of this desire to bring opposing sides together was Feinstein's efforts to save the Headwaters Forest—a 7,500-acre parcel of ancient redwoods, plus twelve smaller stands of redwoods amid 210,000 acres of forest owned by a logging company. The Headwaters was the largest stand of redwoods still in private ownership in the world. The owner, Pacific Lumber Company, had always had very good conservation practices, but when it changed ownership in 1989, the company began to increase its logging. Environmentalists were very concerned that the redwood forest would be jeopardized. During the next ten years, there were ten separate efforts to save the Headwaters Forest, including several attempts at federal legislation, state legislation, and a bond issue. Everything failed. Finally, in 1997, Feinstein stepped in to help broker an agreement between Pacific

Lumber and the federal government to protect this national treasure.

At stake were two competing interests—those of Pacific Lumber, with a substantial number of jobs involved, and those of environmentalists, who were opposed to any logging whatsoever. Over a period of two years, Feinstein pulled the two sides together, even conducting marathon all-day sessions in her office, to reach an agreement that would be both environmentally and economically sound. She secured the funding from the state and federal governments to make it happen.

On the evening of March 1, 1999, after all of these efforts, it initially appeared that the agreement had fallen apart, but Feinstein urged the parties back to the table through late-night conference calls and finally went to sleep around midnight, East Coast time.

She was awakened shortly before 3 A.M.—the absolute deadline to finalize the plan—with a request to get back on the phone. Once more she cajoled all sides to work out their differences and an agreement was reached.

The Headwaters Agreement provided for the federal government and the state of California to acquire the Headwaters Forest, saving the ancient redwoods, some of which are more than two thousand years old. It also required that Pacific Lumber's entire 210,000 acres be governed by the terms and conditions of the Habitat Conservation Plan.

"It was beyond anything I thought we could achieve," Feinstein says. "Of course it didn't please everyone. Many environmentalists wanted the logging stopped completely, and they weren't happy with the compromise. But here's the point: After ten years of failed efforts, we got it done. And as a result, the single largest ancient redwood forest in private hands will be saved for all time."

Women in the Senate are not only redefining the way business is conducted, they're also redefining the meaning of "women's issues." One of the most tenacious stereotypes regarding women politicians is that their venue is the "soft issues" related to family, education, and health. However, when it comes to the "hard issues" of foreign policy, the military, crime, and the budget, voters express more confidence in men. Although women like Madeleine Albright and Janet Reno have blazed trails in domains that were traditionally considered male, neither of them was elected to office, and that makes a difference.

While women often face a credibility gap on the campaign trail, once they're in office that gap disappears. Women serve with distinction on the Senate Armed Services Committee, the Appropriations Committee for defense and foreign operations, and the Judiciary Committee, responsible for international terrorism and narcotics control. Indeed, in many cases, the concerns of local communities are very much bound up in U.S. foreign policy.

Blanche Lincoln has experienced firsthand how devastating it can be to American farmers when the United States imposes trade sanctions on nations like China and Cuba. "Free trade with China is absolutely essential to the agricultural future of this country," she says. "It doesn't make sense to our marketplace to cut farmers off from one-fifth of the world's population. We've been a nation of great prosperity, with a safe and abundant food supply. But our farmers can't compete with farmers in other countries, especially those that are government sponsored."

Lincoln is particularly frustrated with the U.S. sanctions against Cuba, which have a direct impact on the Arkansas rice farmers. "We can ship a higher-quality long-grain rice at a cheaper price than Cuba is currently paying," she says. "Yet we've eliminated ourselves from the market. We'd better start looking at the cost-benefit ratio of sanctions. What have we proved after decades of sanctions? Has Cuba moved closer to democracy? No, but we're losing our farmers."

Late Fall, 1993—In Flight
Between Washington, D.C., and Dallas

Shortly after she came to the Senate, Kay Bailey Hutchison was on a flight home from Washington to Dallas when a man came up to her seat and introduced himself as Larry Joyce. She asked him what he was doing in the nation's capital. His face was etched with a deep sadness. "I was burying my son at Arlington National Cemetery," he told her.

"Did he die in Somalia?" Hutchison asked.

"Yes," Joyce said. "His name was Sergeant Casey Joyce, and he lost his life in his first mission as an army ranger."

Hutchison's heart went out to Joyce. "How can I help you?" she asked.

His voice shaking with emotion, Joyce replied, "I'm a military man. I served two tours in Vietnam. Now I've lost my son, and I can't tell you why. I want to say, 'He died fighting for our freedom,' but that's not what I believe."

Hutchison invited Joyce to testify before the Senate Armed Services Committee, and he remained her inspiration in the ensuing years as she passionately debated America's involvement in peacekeeping missions throughout the world. Hutchison has frequently written and given speeches on U.S. military policy, and she has made seven fact-finding trips to the Balkans. "Larry Joyce died in 1999, but he left an indelible impression on me," she says. "We have allowed ourselves to be pulled into missions that are not necessary for U.S. security. When we set up unending peacekeeping missions in places like Somalia and Bosnia, we jeopardize our ability to respond to real crises elsewhere in the world. This is not an abstract matter. It is very real for thousands of families whose loved ones are proudly and patriotically serving our nation in the armed forces. And it's real for the readiness and security of the United States."

Hutchison employs a personal litmus test whenever the question of U.S. military intervention arises. "I ask myself, Can I look a Larry Joyce in the eye and say, 'Your son or daughter died fighting for our freedom?'"

Redefining and expanding the old labels involves forming unusual alliances, by balancing the needs of each side and stressing the points of commonality.

For example, it has long been believed that the agendas of the business community and environmentalists are at cross-purposes. Environmentalists are opposed to the interests of business, and business is adamantly opposed to the interests of environmentalists.

Senator Barbara Boxer made a place at the table for both the business community and the environmentalists. In sharp contrast to the accepted cant, she believes that there can be no long-term economic growth without a sound environmental policy. The two camps can both achieve their goals—by giving support to each other. And she has demonstrated the result of this unusual synergy through legislation.

The Coastal States Protection Act, which permanently protects California's coast from the potential hazards created by offshore oil drilling, had far-reaching ramifications. Not only did it answer the concerns of environmentalists, but it rescued two of California's most crucial industries: tourism and fishing.

The focus on problem-solving exercised daily by the women of the Senate is a living confirmation of the saying "If you want something done, ask a woman."

9

What We've Learned: Nine Lessons

When the women of the country come in and sit with you . . . I
pledge to you that you will get ability, you will get integrity of
purpose, you will get exalted patriotism, and you will get
unstinted usefulness.

REBECCA LATIMER FELTON
First woman in the United States Senate,
as she took the oath of office

Rebecca Felton spoke those words in 1922, moments after she
took her oath as the first woman United States senator. But she
was referring to a promise that she herself would not be allowed
to fulfill. Felton wouldn't be staying to sit with her colleagues.
Her token service of one day left no time for deliberating the issues, or applying her mind to legislative challenges. However,
since then, the women who have followed in Felton's footsteps
have more than fulfilled her pledge. Their accomplishments are
memorialized in hundreds of pieces of legislation. Individually
and collectively, the women serving in the Senate have created a

legacy through legislation that will significantly improve the lives of Americans for the next hundred years and beyond.

Today, for the first time in our nation's history, women have more than a token presence in the United States Senate. Nine percent falls somewhat short of equity, but it is a far cry from the time, only ten years ago, when Dianne Feinstein declared, "Two percent may be good for milk, but not for representation."

The Senate women want to pass on the lessons they've learned, as a way of encouraging other women to join them, and keep the count rising.

1. When someone says, "Why you?" think, "Why not me?"
Women who win national office are still rare enough to be greeted as newcomers, even when they've served for decades in local and state governments. The female candidate touted as having "come out of nowhere" is likely to have as much experience as her male counterpart, or more. As Barbara Mikulski joked when she won her first Senate race, "I'm a twenty-two-year overnight sensation."

The myth that women who achieve high office are there as if by magic only reinforces the belief of other, similarly inclined women, that they have no right to aspire to such positions. "Our goal," says Olympia Snowe, "is to reach young women and let them know we didn't just get dropped here from the sky. It is our hope that, by telling our individual stories, we can help make it clear to others that it is within their ability to get involved and succeed."

Every woman who has ever blazed a trail in public life has had to overcome a barrage of "can'ts"——*She can't deal with "hard" issues. . . . She can't inspire respect. . . . She can't be tough enough. . . . She can't serve in the Senate and raise children. . . .* Each of the women in the Senate today encountered these and other "can'ts," when she was running for office, including:

"She's too feminine." (Read "soft.")

"She doesn't have national experience." (How many men running for Congress do?)

"She's too strident, too aggressive."

"She hasn't paid her dues."

"She can't raise the money or inspire the support."

"We already have one woman senator." (Incidentally, on opposite coasts of the country, in Maine and California, a curious coincidence occurred. These states have placed two women in the Senate to represent them. The objection sometimes given to a woman running—the specious argument that "we already have one"—has already been overcome. "And," adds Barbara Boxer with delight, "in California, many said in 1992 that we could elect one woman, not two and certainly not two Jewish women!" Often during the campaign, Boxer noted that "nobody made an issue over electing two Protestant men.")

"Our state has never had a woman senator."

"She's too young."

"Who will watch the children?"

If there is any doubt that the subtext is "She's a woman," one need only consider the same criticisms leveled at a male candidate:

"He's too manly."

"He's too aggressive."

"He hasn't paid his dues." (After several terms in the state legislature, or Congress.)

"There's already a man serving from our state."

"He can't command the backing."

"We've never sent a man to the Senate from this state."

"He's too young."

"Who'll take care of the children?"

If these arguments seem comic when applied to male candidates, why aren't they just as comic when applied to women?

"The real message," says Dianne Feinstein, "is that women in office can do everything that men can do. There is no reason they shouldn't be given the same respect when they decide to run for office."

Former congresswoman and vice-presidential candidate Geraldine Ferraro, who made two unsuccessful runs for the United States Senate from New York, was once asked why she would run a second time after having already been defeated. She replied, "The fact that it is a struggle is never a good enough reason not to run. You do it because you believe you can make a difference. You do it because it's an opportunity available to you that could barely have been imagined by your ancestors. You certainly don't do it for money or power—which is the cynical interpretation many people have of politicians and their motivations. You can go broke running for political office, and as for power, well, it's a limited and fleeting thing. An effective representative takes her power from her constituents. The reason you run is that your vocation is public life."

But often the greatest "can't" comes from inside. The key, Susan Collins advises, is to quiet the inner critic. "Women always say, 'I'm not ready' . . . 'I need more experience.' They put off deciding. They need to look at what they've already accomplished, and see that they can do it."

2. *Remember who you are, and where you came from.*

When Barbara Boxer took the oath of office for the United States Senate, a reporter asked her if she was going to change her style. "No," she said, "I'm just going to be me." She found the question puzzling. The voters elected *her*, not someone else. Why would she change? There is a widespread perception that service in Washington changes people—as if there were something in the water. "I suppose it could go to your head," Patty Murray muses. "People treat you with deference, they address you as 'Senator,' no matter how you introduce yourself, and members have their own private elevators. But I constantly remind myself why I'm here."

Mary Landrieu, who once dreamed of becoming a social worker, says that she has never lost the desire to do good for others. "I like having power," she acknowledges. "But it's not power for power's sake. Power shouldn't be something you want because it lifts your ego. It's power to do more good than you can do without it."

"For me, the whole idea of public service was to work on problems," Olympia Snowe says. "When I first came to Congress, it seemed that there was so much pomp and circumstance. I thought it was funny. I just thought, 'Give me the work.'" Even if Snowe had been tempted to get caught up in the importance of her position, the constituents would have set her straight—especially the children. She laughs when she recalls the thank-you notes she received from students after visiting their elementary-school classroom. "One of them said, 'Thank you, Mrs. Snowe, for visiting our class. It was one of the two best parts of the year. The other is when we dissected cow lungs. Fortunately, they never told me which one ranked first.'"

Blanche Lincoln only has to walk into her home to be reminded of who she is. Her energetic four-year-old twins don't stand on ceremony. In fact, when she was being sworn in to the Senate, the voice of one of her sons pierced the solemnity of the occasion with a loud cry: "That's Mommy!" Lincoln is always surprised when the people back home make a fuss over her. "The pull to be a celebrity is incredible. You have to understand that it's just a part of the job, downplay your high visibility, and show people that you're there to add value."

Remembering who you are is also key to building your credibility on important issues. Dianne Feinstein has been most effective in doing this with the gun-control issue. In 1993, less than a year into her Senate term, she made a strong argument on the Senate floor for the passage of the assault weapons bill. Idaho Republican Larry Craig challenged Feinstein's facts, and added condescendingly, "The gentlelady from California needs to become a little more familiar with firearms and their deadly characteristics."

Feinstein shot back, "I am quite familiar with firearms. I became mayor as a product of assassination. I found my assassinated colleague and put a finger through a bullet hole trying to get a pulse. I was trained in the shooting of a firearm when I had terrorist attacks, with a bomb at my house, when my husband was dying, when I had windows shot out. Senator, I know something about what firearms can do."

There wasn't a sound on the Senate floor when Feinstein was finished. Her amendment passed 51 to 49.

3. Create a team effort.

Studies have shown that men and women often have different styles of leadership. Men tend to lead through power and autonomy. Women are more inclined to lead by inspiring others to join the process. They understand that relationship building is central to building support. "Americans see us fighting on the evening news," Blanche Lincoln says, "and they think that's what it means to be a senator—you fight and make speeches. People don't see our caucus work on TV. They don't see the ways we work for bipartisan solutions. But for me, that's the real excitement of being in the Senate."

4. Don't take it personally—and don't make it personal.

Barbara Mikulski has an unwavering rule: stay focused on the goal. "If you look at the negative stereotype of women, it's said that we personalize everything. But I never make anybody the issue. I always keep the *issue* the issue so I can keep working in the Senate."

After Mary Landrieu endured nearly a year-long investigation into her Senate campaign by Republicans on the Rules Committee, she was tired and certainly frustrated. But she didn't feel bitter. "I had a choice," she says. "I could decide that the Republican leadership and the members of the Rules Committee were my enemies—which they weren't; they were just doing their jobs—or I could put it behind me and do the work I was sent to do. I chose the latter, not only because it was the right thing to do, but because it makes better sense to make friends than to make enemies."

5. Identify the felt need.

Don't say, "I want to be a politician." A politician is an empty suit. There is nothing inside. Nor is it enough to say, "Vote for me because I'm a woman." When Blanche Lincoln ran for the

Senate in 1998, she was criticized in some circles for being too gender neutral—for not placing more focus on women. She defended her position, saying, "I focused on the issues that I knew were important to families and people of both genders. They also happened to be important to women, but the point was, people wanted to hear about the issues."

The women agree that the key to actually *having* an impact—as opposed to *talking* about having an impact—is to do the fine detail work involved in targeting an issue that is truly felt by people on the grass-roots level. "There are plenty of excellent ideas floating around that are top-down ideas," Barbara Mikulski says. "But until an idea becomes a felt need at the grass roots, it won't take hold. A felt need has to be experienced both emotionally and intellectually by a large enough group of people." Only then, Mikulski advises, can you begin the process of figuring out how to respond.

"To be successful in representing people, you should be driven by a cause or a principle, not because you want to be in politics," says Kay Bailey Hutchison.

6. *Respect your losses.*

In 1990, Dianne Feinstein ran for and lost the race for governor of California by a narrow margin. Two years later, when she ran for the United States Senate, she won. In the Senate, Feinstein says, she found her calling—the work she was meant to do. So, in 1996, when she was approached about running in the upcoming gubernatorial election, she gave it serious consideration, then declined. "I had found what I was meant to do," she says simply.

Susan Collins, who also lost a race for governor of Maine before she ran for the Senate, can look back on that defeat as an important preparation ground for her future campaigns. "It was a very difficult but very exciting race. You learn a lot of lessons from it. You learn to keep going no matter what, no matter how hard it is. In some ways it is a real test of endurance and perseverance and commitment. Because you can't rest, you can't give up, no matter what's happening. You've got to keep going."

Barbara Boxer remembers that the first time she lost an election—for supervisor in 1972—it was a terrible blow. "I could hardly look at myself in the mirror," she remembers. "But I've learned how to take it on the chin." During the toughest part of the Senate campaign in 1992, her two adult children presented her with a Dr. Seuss book, *Oh, the Places You'll Go!*—underscoring one passage in particular:

> *Wherever you go,*
> *you'll top all the rest.*
> *Except when you don't.*
> *Because sometimes you won't.*

"I keep that phrase from Dr. Seuss in mind whenever I lose a vote, or have trouble moving a piece of legislation along," Boxer says. "It helps me focus on what I can do, not what I fail to do. It keeps me coming back to try again."

Olympia Snowe learned that lesson in 1996, when she successfully overturned a proposed cut in student loans. "I come from a working-class family, and my education was made possible through student loans. So I knew the cuts would have had a devastating impact." Snowe and fellow Republican senator

Spencer Abraham, of Michigan, offered an amendment to restore $6.3 billion to the program. The Snowe-Abraham Amendment lost, but Snowe felt the issue was too important to give up. So she worked with Senator Paul Simon (D-Illinois) to introduce a bipartisan amendment that restored even more funding: $9.4 billion. The Snowe-Simon Amendment not only passed but did so with an overwhelming vote of 67 to 32.

"Our experience shows that when you're playing political football on the floor of the Senate, you have maintain control and possession of the ball," Snowe says. "Our amendment gave prospects for bipartisanship new hope, but more important, it gave needy students in America a second chance."

7. *Control your agenda.*

The women senators acknowledge that there are tremendous opportunities in the Senate—and that women can bring to the table a unique way of using their power to achieve their goals.

Blanche Lincoln recalls that when she arrived in Washington to start serving in the Senate in early 1999, President Clinton invited her to the White House to watch a movie. "He was making a gesture to a fellow Arkansan," Lincoln says. "I brought my list of agenda items that I wanted to discuss with him, and when I handed it to him, I realized I'd made a faux pas. He gave me a funny smile, and put the list in his pocket. I was so embarrassed, thinking, 'Oh, he really meant we were going to watch a movie, and I blunder in with my legislative priorities.' But to me it just made sense that if you're going to spend a couple of hours with the president, you might as well take advantage of the proximity."

Barbara Boxer points out that controlling your agenda can also mean waiting and listening. "You can't come to the Senate and announce, 'Well, I'm here and I'm wonderful,'" she laughs, brushing aside the notion that she rode into the Capitol on a white horse. "When I first got here, it was a little bit like a boxing match. People were kind of circling the ring, looking at me, wondering when I was going to throw my first punch—and if it would be at them. But I wasn't thinking about that at all. Sure, I was passionate about certain issues, but I realized how important it was to listen and learn the lay of the land. You don't come to the Senate and declare that you're an expert, because there are people who have been here a lot longer. You have to listen. Not that you don't speak—of course, you do—but you listen carefully and find a place for yourself."

8. *Ignore the babble.*

Every woman in government hears a constant background cacophony of observations and criticisms that have nothing whatsoever to do with her ability to govern. Much of this babble feeds on negative stereotypes, and the senators have all learned to avoid getting sidetracked by them.

"Others will always be trying to define you by their standards," Kay Bailey Hutchison observes. "You just can't pay attention." As an example, she points to a political appearance by the actress Cybill Shepherd during her campaign, where Shepherd said that Hutchison was "no good for women and children." Hutchison didn't pay any attention. "The idea that because someone is a famous star, they have more important opinions than other people, is ludicrous," she says. "She's welcome to her opinion, but I have a record. The homemaker IRA and the antistalk-ing bills are good for women and children."

9. *Pass it on.*

Change takes time. As Patty Murray has joked, "Some pundits still think change in the Senate means brightly colored dresses in a sea of gray suits." But the true task of passing on the opportu-nity—of reaching out to the next generation of women so that they can build on your success—is a long, gradual process.

Over the years, Susan Collins has frequently been invited to speak at Maine's Girls' State Convention. It is a special opportu-nity for young women across the state to participate in a week of mock government—including elections, debates, and problem solving. In 1970, Collins herself attended as a high schooler, and when she speaks to each new crop of potential public servants, she communicates the joy she experiences from public service—and she challenges them to reach high and consider making that life choice themselves.

Mary Landrieu regularly shares her experiences and ideas with groups of schoolchildren, who communicate through scheduled Internet chats. "They'll ask everything from, 'How can you be a mom and do your job?' to 'What legislation are you working on?'" Landrieu says. "They give me their opinions, too. I learn a lot from them, and it helps the children feel connected and ex-cited about government." When Landrieu came to the Senate, she organized a program for the annual Take Our Daughters to Work day. She sees it as her way of passing the torch—reaching out to girls at a young age and inspiring them with their possi-bilities.

Patty Murray has organized a Senate Advisory Youth Involvement Team (SAY IT!) across Washington State that she consults with frequently. "I want them to see that you can do great things as a policymaker or a legislator," she says. "I value the input of young people. But I also want to connect them with what I'm doing. I want to raise a generation of kids who recognize that politics is something that affects their life, and rather than sitting back and letting those policies drive *them*—*they* can drive those policies. I'm trying to empower them with that knowledge."

There is also an appreciation of the power of symbols to influence the way young women perceive their opportunities. When Olympia Snowe successfully lobbied for the commissioning of a portrait of former Maine senator Margaret Chase Smith to hang in the Capitol, she noted that not only was it a richly deserved honor, but "the portrait of this woman, who served with such dignity and honor, is a testimony to the possibilities that exist for all women."

In 1999, on the eighty-seventh anniversary of the founding of the Girl Scouts of America, Kay Bailey Hutchison and Barbara Mikulski cosponsored a resolution granting the organization a federal charter—creating a permanent link between the work of the Girl Scouts and Congress. The resolution passed unanimously, making the Girl Scouts the first organization for girls to be granted a federal charter by Congress.

When Blanche Lincoln was first elected to Congress at the age of thirty-three, she recalls, several of the older congressmen just couldn't get over it. "Why, I have granddaughters who are your age!" one of them exclaimed. Lincoln flashed a big smile, and replied, "I'll bet your granddaughters are glad I'm here."

———

Nearly eighty years ago, when Rebecca Latimer Felton stood on the floor of the Senate and pronounced her vision for women, it was a breakthrough that she was even allowed to stand there for one day. Today, every woman who feels the call to serve is encouraged to get involved. The women of the United States Senate have a message: "There's work to be done, and we need *you* to do it."

AFTERWORD

. . . And Then There Were Thirteen

Every Tom, Dick, and Harry is now every Maria, Jean,
Hillary, and Debbie.

BARBARA MIKULSKI

December 6, 2000—The Hart Senate Office Building

Thirteen glistening white china cups surround an elegant silver
coffee service, artfully arranged on a side table in Senator Barbara Mikulski's office. Maryland's Democratic senator is hosting
a "power coffee" for her women colleagues in order to welcome
four new additions—two-term Michigan congresswoman Debbie Stabenow, newly elected to the Senate; Washington State's
Maria Cantwell, a former congresswoman and high-tech business executive; Hillary Clinton, the nation's First Lady and New
York's senator-elect; and Missouri's Jean Carnahan, appointed
only two days earlier to fill a seat won posthumously by her husband, Governor Mel Carnahan. Today's coffee will be a biparti-

san show of support by the nine women senators, who are delighted that their ranks have grown. Earlier, when Mikulski was asked by a reporter, "How do the new senators feel about joining an institution that is still an old boys' club?" she was quick to dispute the characterization. "Not quite *still* an old boys' club," she countered. "Women are thirteen percent of the United States Senate. And we're still counting."

Mikulski clearly relishes her role as Dean of Senate Women. Since 1992—when, for the first time in our nation's history, the ranks of women in the Senate rose to a number greater than two—Barbara Mikulski has provided both philosophical and strategic guidance for new women members. Democratic and Republican women alike have benefited from her wisdom.

In January 2001, the four women will be sworn in to a Senate that is evenly divided along party lines. At the time of this coffee, the outcome of the presidential election remains unclear. The recount debate wages on in Florida, and the Republican candidate, George W. Bush, will not be certified as the eventual winner and president-elect until later in the month. Yet as the women arrive at Senator Mikulski's office, there is no sign of the partisan tensions that have sprung from this current uncertainty. They greet one another with relaxed good spirit, and spontaneous cheers of welcome are offered to the new members. As the senators take their seats in a semicircle around Mikulski, the prevailing mood is one of camaraderie. "Welcome to the most exclusive club in Washington," offers one senator, and there is appreciative laughter. The most exclusive club in Washington was until recently a preserve unswervingly devoted to men. No more.

Like the women before them, each of the new members will

take her seat having overcome tremendous barriers, and there are several "firsts" to celebrate. Jean Carnahan will be the first woman to serve in the Senate from Missouri, and many would agree that she has paid the highest price. "This is a bittersweet moment," Carnahan admits quietly. "Everyone else is here because of their win. I'm here because of my loss."

It is a loss that few can fathom. Carnahan is in Washington only two days after receiving her appointment to the seat won posthumously by her late husband, Governor Mel Carnahan. He and the Carnahans' eldest son, Randy, died on October 16 when their small plane crashed as they headed for a campaign appearance. Chris Sifford, a top aide, also perished. Yet in spite of her loss, Jean Carnahan is here, ready to work, and the admiration for her strength and courage is palpable everywhere she goes.

Hillary Rodham Clinton has broken the most barriers. The only First Lady ever to be elected to the Senate, she is also the first woman senator from the state of New York. After she takes her oath on January 3, she will be in the unique position of holding both titles until January 20. Her husband, the president of the United States, had wanted to attend today's luncheon for Senate spouses, but he reluctantly bowed out because of the added security his visit would necessitate.

There are other breakthroughs. Debbie Stabenow is the first woman to be elected to the Senate from Michigan. With Maria Cantwell's election, Washington becomes the third state to be represented by two women.

The press is in full force today, crowded into the narrow corridor outside Mikulski's office. They've been promised a photo op, but first there's work to be done. The door to Mikulski's office is closed, and the women get down to business. This is, after all, a *power coffee*, not a coffee klatch.

"I love being a United States senator," Mikulski tells them, "because I get to fight for the day-to-day needs of my constituents and the long-range needs of the nation." To do that effectively, she says, requires power, and power in the Senate is achieved through careful attention to the three Cs: "Connecting with our *colleagues*, getting on the right *committees*, and staying in touch with our *constituents*—those are the keys to success in the Senate," Mikulski says.

Their most important obligation, Mikulski reminds them, may find its completion on the Senate floor, but its foundation is laid and its scaffold erected from the practical organization of offices, phone lines, mail rooms, scheduling, staff—the minutiae of their operations. "Every one of us came here with a passion for the issues," she says. "We all have good intentions, but we have to *operationalize* our good intentions."

Coffee has been served, wisdom has been shared. After a while, the clutch of avid reporters and photographers who've been clogging the hallway outside Mikulski's office are allowed in to capture the event for the next day's newspapers.

"Is thirteen enough?" a reporter jovially calls out to Mikulski.

Mikulski's eyes dance with good humor. "Only thirty-eight more women and we'll have a majority," she quips.

"Mrs. Clinton, have you been surprised by anything?" another asks. "*Everything*," she replies with a tired laugh. It's day two of a three-day orientation, and the freshmen have been immersed in caucuses, workshops, lectures, and coffees since early the previous day.

The next question evokes smiles and chuckles, its very expression speaking volumes about how far women have come. "Why can't a male senator do everything a woman senator can do?" a reporter asks.

Kay Bailey Hutchison graciously tackles the reply. "Sometimes, from our experience, there are issues that men just haven't thought about," she explains. "Like the Homemaker IRA that Senator Mikulski and I sponsored, or the women's health issues we've worked on together. These concerns come up in women's lives, and we bring them to the attention of the entire Senate. Most of the time our male colleagues are supportive once we've made the case."

A reporter wants to know if the women are worried that their collegiality will be strained by a divided Congress.

"We check our party hats at the door," Mikulski says firmly. "In this fifty-fifty Senate, we're asking, 'How can we set the tone of civility?'" Turning to Senator Hutchison, she adds, "You and I disagree on taxes, among other things."

Hutchison nods. "Yes, but the key is that we don't expect everybody to be the same. We can disagree and still be supportive of one another."

This isn't just talk. They mean it. In July, when the nine women senators appeared on *Larry King Live* to talk about their book, they stunned their host and many of the program's viewers by stating that they would not campaign against other Senate women. That decision, Mikulski explained, was forged from a bitter experience. "When Olympia was running the first time,

George Mitchell asked us to go up [to Maine and oppose her]. And I will tell you, it was one of the most melancholy things I ever did. I apologize to Olympia."

King looked incredulous. "For campaigning against her?"

"Yes," Mikulski replied, "because this is a great woman. I think we all learned a lesson, which is that we are not going to campaign against each other. We are going to duke it out in the Senate. We have different issues, different parties. But I think we all feel that every one of us has made a difference, and we want to support that. When we have been together, we have brought about change. And we are proud of each other."

In a season of fierce partisanship and political thunder, the women of the Senate have deliberately concentrated on their similarities, not their differences. The morning after the election, Olympia Snowe placed a call to Hillary Clinton, offering her congratulations. The two women laughingly recalled their first meeting back in 1989, at the National Governors' Association conference for spouses. "Now our husbands are the spouses," Snowe chuckled.

Kay Bailey Hutchison, the most conservative of the Republican women, and an outspoken critic of the Clinton White House on many occasions, was equally welcoming to Hillary Clinton. "Her page is blank and she is going to be able to write on it," Hutchison said. "And she will be able to write the kind of senator she is going to be, and she will be accepted for what she chooses her role to be."

If it were only a matter of courtesy, the women's relationships might not be all that unusual. The Senate, after all, is a place of great civility—at least on the surface. But there is more than just the appearance of civility in the room. There is a real empathy, a bond that grows from shared experience. As the women prepare to leave Mikulski's office, they make plans to get together soon for their regular informal dinner. Feinstein offers to host the occasion at her Washington home.

"Maybe I'll make my famous paella," she offers.

Mikulski laughs. "Dianne, we've been hearing about that paella for eight years, and we've yet to sample it."

Revealed is yet another great unspoken truth. When you're doing the work of the people, who has time to cook?

While it is a remarkable achievement for a woman to win a seat in the United States Senate, it is a validation of her success in that position when she wins reelection. Of the nine women serving in the year 2000, three were up for reelection, and they were all returned to the Senate with resounding victories. In Texas, California, and Maine, voters went to the polls, often crossing party lines, to show approval for the job their women senators are doing.

In Texas, Kay Bailey Hutchison won a second full term with 65 percent of the vote, easily defeating Democrat Gene Kelly and setting a record of her own. With more than four million Texans voting for her, she now holds the position of top vote-getter in the state's history. "The first time I ran for the Senate, it was so tough," Hutchison recalls. "This time I had the credibility because of the work I had done, and I was treated as the favored incumbent in every way."

Hutchison has also gained the respect and confidence of her

colleagues in the Senate. In December, she was elected without opposition to the position of vice chairwoman of the Republican Conference. She is the first Republican woman to be elected to a Senate leadership position since Margaret Chase Smith served as conference secretary in the 1960s. "It really does make a difference in what you are able to accomplish," Hutchison says of her leadership role. "You're in the room when there are discussions about strategies and priorities. Your voice is heard."

In California, Dianne Feinstein handily beat back San Jose Republican congressman Tom Campbell to win reelection with 55.9 percent of the total vote, 1.4 million more than George W. Bush and 100,000 more than Al Gore received in the state. Her total was the most votes cast for a senator in United States history. As cochair with her California colleague, Senator Barbara Boxer, at the Democratic National Convention in Los Angeles, Feinstein was a visible presence on both the state and national levels as a passionate advocate for gun control.

Speaking to the convention, Feinstein called for the nation to make family safety a priority. "Do we really think sensible gun laws, safe children, and secure schools are too much to ask for?" she asked. "For a single family in California, a million moms, or every American . . . the answer is, it's never too much. We will never stop fighting—not until every street and every school, and every child in every community, is safe once and for all."

Briefly derailed by a broken leg just after Labor Day, Feinstein refused to let the injury sidetrack her campaign. Instead, she ran an upbeat race, reinforcing commitment to her signature issues—gun control, victims' rights, HMO reform, education, and the environment.

In Maine, Olympia Snowe was reelected with 69 percent of the vote, winning a decisive victory over State Senate president Mark Lawrence. Lawrence was required to leave office as a result of Maine's strict term-limit laws, and he was a credible and well-funded opponent. His campaign failed to gain traction, however, because he targeted issues, such as prescription drug coverage and school construction, that Snowe had long supported. With her reelection, Snowe holds the record (currently standing at ten victories and no defeats) for consecutive election wins in the state of Maine, no small achievement in the home of prestigious former officials such as George Mitchell, William Cohen, and Margaret Chase Smith.

In her reelection campaign, Snowe championed the needs of struggling working families, committing to fight to procure affordable prescription drug coverage and to find ways of meeting the needs of the uninsured. She will be in a unique position to fulfill that promise, having secured a seat on the Senate Finance Committee, which oversees about two-thirds of the federal budget. That, combined with her place on the Budget Committee, will put Snowe in a crucial position to help shape federal spending priorities.

Snowe believes her strong showing in the 2000 election is also a sign that the people of Maine support her leadership in building coalitions across party lines to achieve results. "People want independent-minded officials who are willing to cross party lines," she says. "They want representatives who make things work, who build consensus and compromise."

The powerful twofold message of the 2000 election is this: Women can get elected to national public office. And women can produce the results required to stay there. Already impressed with the stamina and achievements of the original nine, in the early months of the 107th Congress the nation will be introduced to the four unique, dedicated individuals who have elevated the number of women in the Senate to thirteen.

February 7, 2000—Purchase, New York

New Yorkers are a tough crowd. They are challenging, wary, hard to win over. Their truth meters are finely honed to detect insincerity, especially from politicians. But as Hillary Rodham Clinton stood in front of a crowded gymnasium at the State University of New York at Purchase and officially announced her candidacy for the United States Senate, the comfort level in the room was high. After nearly a year spent absorbing the spirit and issues that defined the state, Clinton had earned the respect of even some of her fiercest critics. Still ahead were nine arduous months of nonstop campaigning, but one thing was clear that day: This was not the celebrity candidate, the First Lady of the nation dabbling in a political run. Clinton was a fighter. When she promised, "I will fight for you," she really meant it.

As she began traveling the state in early 1999, listening to people's concerns and familiarizing herself with their neighborhoods, she was quietly disarming some of New York's most cautious citizens. In school lunchrooms, libraries, town halls, senior centers, and factories, she listened and took copious notes. She stayed behind after events to shake every hand. She lingered to talk to individual parents about their children, to workers about their jobs, to the elderly about the high cost of their prescriptions. She had an instinct for listening, and it delighted and sometimes surprised people. After an event for teachers and parents in a small upstate town, several people marveled that Clinton hadn't seemed to be in any hurry to leave. She spoke personally to everyone in the room, sharing her own experiences as a parent and empathizing with people's concerns. Observing that Clinton was genuinely interested in what she had to say, one mother said, "She listened with both ears."

Those grass-roots reactions weren't always reflected in the media. As she traveled the state, Clinton was subjected to a constant cacophony of querulous press voices, like a Greek chorus chanting, "What right do you have? What have you ever done?"

In any other context, these would have been absurd questions. Although she had never personally held office, Clinton had been in public service for most of her adult life. Her partnership with her husband, first in the Arkansas governor's mansion, then in the White House, had provided her with opportunities far beyond the scope of most politicians. She was an activist First Lady, a pioneer in health care, and an internationally renowned advocate for women and children. It was doubtful that anyone honestly thought she wasn't up to the job. In New York's Senate campaign, the real question was related to birthright: Could a woman born and bred in the Midwest, who had then spent most of her adult life in the South, convince the citizens of New York State that she was the senator who could best represent them in Washington, D.C.?

In many respects, Clinton represented the quintessential baby boomer, part of a mobile generation that had effectively expanded its local and regional borders. No place in the country reflected that reality more than New York. Its cities and suburbs teemed with outsiders who had initially come from elsewhere and now called New York their home. The rich fabric of its population was woven not only from the constant influx of immigrant populations seeking a better life in America, but from countless transplants who had chosen to leave their home states and live in New York. Nonetheless, whether they were native New Yorkers or had adopted the state later in life, its citizens were remarkably cohesive. They spoke the common language of New Yorkers, and could spot a tourist or an impostor a mile away.

As Clinton stood on the stage that February 7, she spoke from the heart, addressing the two thousand supporters in the room and the vast television audience beyond. "Some people are asking why I'm doing this here and now," she said. "That's a fair question. Here's my answer—and why I hope you'll put me to work for you: I may be new to the neighborhood. But I'm not new to your concerns. . . . For over thirty years, in many different ways, I've seen firsthand the kind of challenges New Yorkers face today. I care about the same issues you do. I understand them. I know I can make progress on them."

From the sidelines, Rudy Giuliani, mayor of New York City and the Republican Party's designated candidate, gleefully gibed at Clinton's efforts to establish her credibility in New York. Appearing before reporters, decked out in a Yankees baseball cap and jacket, his tone was sarcastic. "She says she was a Cubs fan all her life," he said with a smirk. "Now she says she's a Yankees fan. Which is it?" Perhaps Giuliani believed he could defeat her by poking fun at her from his office in City Hall, but Clinton simply refused to be drawn in. From the outset she had decided that if she was going to make a run, she was going to do it on her own terms and set her own agenda. So while Giuliani remained at his post, delaying an announcement of his candidacy or a debate on the issues, Clinton kept working. She was the energizer candidate, earnestly setting out to visit every one of New York's sixty-two counties and to study every issue of concern to its citizens. She familiarized herself with the local landscapes—from Ithaca, with its stunning waterfalls and lakes to the elevated vistas of Essex County in the Adirondacks to the fertile cradle of the Hudson Valley to the man-made spires of Manhattan Island to the sandy oceanfront on the easternmost point of Long Island.

In her bestselling book *It Takes a Village: And Other Lessons Children Teach Us*, Clinton wrote, "There's an old saying I love: You can't roll up your sleeves and get to work if you're still wringing your hands." That philosophy infused every moment of Clinton's campaign. Even Republican partisans acknowledged a grudging admiration for how hard she worked. One prominent Republican political strategist admitted, "It kills me to say this, but it's been a pretty much flawless campaign."

By the early summer of 2000, with only a scant few months remaining until the election, Republican Party leaders began openly expressing their concerns about Rudy Giuliani. Although the mayor of New York City was sitting on a substantial war chest, he had yet to launch a real campaign. He seemed almost indifferent toward what was supposed to be the political battle of his life.

The reason soon made itself clear. In late May, the mayor announced that he had been diagnosed with prostate cancer and would not make the run.

New York State Republicans scrambled to come up with a replacement, and quickly settled on Rick Lazio, a telegenic four-term congressman from Brightwaters, Long Island. Lazio had the look of a winner. He was smiling, breezy, and affable, a hometown boy made good. Lazio immediately positioned himself as the un-Hillary, and he stayed with that theme. An early campaign mailing, sent to Republicans across the country, read: "Six words will tell you all you need to know about my candidacy. *I'm running against Hillary Rodham Clinton.*" The cash poured in.

Lazio continued this strategy until the bitter end. The flaw in his approach eventually became clear. He was already preaching to the converted—the hard core of conservative New Yorkers who would not vote for Clinton under any circumstances. Rick Lazio never succeeded in expanding his base of support—or in making a case for why voters should choose *him*, apart from the fact that he wasn't *her*.

On election day, Clinton held a slight lead in the polls, and the pundits predicted that she would squeak by with a very narrow victory. The pundits *were* right about one thing—Hillary Rodham Clinton won. But nobody predicted that her margin of victory would be so great. The final tally gave her an edge of nine percentage points. Even more surprising, she carried a number of upstate counties that had previously been considered steadfast Republican strongholds.

Appearing before a cheering throng of supporters, with President Clinton and daughter Chelsea by her side, Clinton expressed her gratitude. "Sixty-two counties, sixteen months, three debates, two opponents, and six black pantsuits later, because of you, here we are," Clinton exultantly recounted. "You came out and said issues and ideals matter. Jobs matter, downstate and upstate. Health care matters. Education matters. The environment matters. Social Security matters. A woman's right to choose matters. It all matters, and I just want to say, from the bottom of my heart, thank you, New York."

Life with Hillary Rodham Clinton these days resembles a minimarathon, her staff moving at double time to keep pace with their new boss. Her early weeks as the United States senator from New York involve a series of sprints between committee hearings, constituent meetings, and the Senate floor. Time is short, and the echoing corridors are very long. In the Dirksen Building, which houses the committee hearing rooms, as well as Clinton's temporary basement office, a major renovation project has shut down some of the elevators and a fine layer of plaster dust clings to every crevice in the stairwells. The basement hallways are piled high with displaced office equipment, desks, and chairs, adding to the air of general disarray.

On a recent morning, Clinton's official activities require a deft juggling act to accommodate multiple committee hearings. But the first order of business is a speech on the Senate floor, in which she sounds a warning about President Bush's proposed tax cut. "I know and respect that President Bush supports faith-based programs," she states, "but his tax plan should not be one of them. Going ahead with a huge tax proposal now is like getting a letter from Ed McMahon and going out to buy a yacht."

Following her statement, Clinton heads back to the Dirksen Building to attend a packed hearing of the Senate Committee on Health, Education, Labor and Pensions, where Secretary of Education Roderick Paige is answering questions and responding to concerns about President Bush's education proposals. From there it's up two flights to a Budget Committee hearing on Medicare reform and prescription drugs.

Between destinations Clinton is trailed by a flock of high school students hoping for a word or a handshake. She invariably gives them more. As she trots down the corridor, she raises a hand to beckon the students forward. "Come on up," she calls, then stops to chat and have her picture taken several times with different cameras. The kids are by turns thrilled and incredulous. As Clinton disappears into the elevator, two teenage girls slide to the floor clutching their cameras and each other, in the throes of an ecstasy usually reserved for the sighting of rock stars.

Clinton is gracious and cheerful whenever people approach her, but it is in the committee hearings that her face really lights up. When this is pointed out to her, she gleefully confesses, "I plead guilty to the charge of being a policy wonk." She continues, "Policy matters. Take the earned income tax credit, which few people understand but which lifts millions of children and families out of poverty. It really matters. It also matters that we devise a prescription drug benefit plan. I meet with seniors from all over New York. They come up and show me their prescription drug bills, and they're *trembling*—literally being made sick with worry. I love the idea that I can be part of developing policy to solve these problems."

Clinton clearly enjoys the job, and she can even find humor in the ceaseless silly speculations that surround women in public office. These speculations center not so much on ideas and achievements as on hair and clothes. Clinton can tell a story with flair, and she regales her staff with a classic example of the special standards that seem to apply to women. "One of my Republican colleagues came up to me on the Senate floor. He said, 'I don't know how you put up with this. I was just chased through the subway by a reporter, and he finally caught up with me, panting, 'Senator! Senator! I have a *really* important question to ask you.' And I said, 'All right, what is it?' And he said, 'Senator, what do you think of Mrs. Clinton's *hair*?'"

Clinton knows that women are often held to different standards than men when they seek public office. She believes that her own success in overcoming the barriers imposed by these standards is a testament to what women from every background and every walk of life can achieve.

HILLARY RODHAM CLINTON

There was certainly a special set of circumstances facing me when I decided to run for office. But, in general, I think the same kinds of challenges confront any woman seeking office in our system. We have the extraordinary task of presenting ourselves to the public in a way that is true to who we are—and many of us are not only active publicly, but we're mothers and wives and daughters and grandmothers and friends. We have all of these other obligations. Trying to convey who we are as whole people—not just a sound bite version—is still a greater challenge for women than it is for men.

In my Senate race, we just broke down another barrier. I said, "I've had a full life doing many different things—my profession, philanthropic endeavors, political work. I want to be judged on who I am and what I've done and what I plan to do." I was fortunate enough to convince the majority of voters to allow me to do that.

Contrary to what many people might believe, New York was a very tough place to run as a woman. Before I ran for the Senate, a woman had never won a statewide race on her own. There were tremendous impediments that became apparent to me as I began to focus on the challenges I faced. I had to overcome those challenges one by one. It was hard, but I've always cared about the world around me. I deeply love our country. My message to young women is that, as tough as the political environment is, if you care about making a difference, you have to be willing to get out there and try. There's no set rule for how you prepare. You can come at it from many different perspectives. Some women get their education and establish careers before seeking office. Some raise their children first. It's a wonderful testament to the options available to women. We've broken through all of these barriers so that individual women can make the choice that's right for them.

Choosing public service is a very personal decision and no one can make it for you. You have to look deep inside yourself and examine your motives, and think hard about what you would do to make a difference. Then you have to be willing to subject yourself to a very tough and sometimes mean-spirited political environment.

But I can tell you, from the heart, it's worth it.

February 14, 2001—Russell Senate Office Building

Debbie Stabenow has placed a large framed picture of a young girl at the center of her Senate office desk, set apart from a cluster of family photographs. The girl's radiant smile lights up her face and captivates visitors. Her name is Jessica Baccus, and she is a powerful emotional symbol of the commitment Stabenow has made to the people of Michigan.

Jessica Baccus had a rare metabolic disorder that required intensive treatment. When her HMO changed, it sent her parents into a tragic spiral, forcing them to pay for the care she needed out of their own pocket while spending hour after hour on the phone with insurance company bureaucrats, explaining their situation. On September 10, 1999, Jessica passed away. In the final days of her life, her mother will always regret, she couldn't be on the front porch blowing bubbles and reading books to Jessica; she was instead fighting the insurance company bureaucracy to get her daughter the treatment she needed.

Stabenow is still moved when she talks about what happened to Jessica. "This family had insurance," she says passionately. "They did everything right. What happened to them is what every working family in America fears—that when they really need the insurance they've been paying for all their lives, it won't be there for them. I promised the people of Michigan that Jessica's picture would stay front and center on my desk until we pass the Patients' Bill of Rights."

Stabenow's empathy for the hardworking people of her state shines through as she speaks. Another focal point of her campaign was her pledge to help lower the price of prescription drugs for senior citizens. "So many of our senior citizens are forced to choose between the medications they need to stay alive and basic necessities like putting food on their table or heating their homes. It's a disgrace!" she says. "We're America! We can do better than this!" Stabenow is particularly incensed that drugs bought in the United States at the manufacturer's inflated prices can be purchased at a much cheaper cost in Canada. During the campaign, she made news by chartering a bus and taking a group of senior citizens across the U.S.-Canadian border to Windsor, Ontario, to fill their prescriptions for about half the price they must pay in the United States.

Wasting no time, Stabenow spent her first week in office making good on key campaign promises. She cosponsored a bill that would place medical decisions back in the hands of doctors and patients and hold insurers accountable when they make decisions that hurt patients. She also introduced the Medication Equality and Drug Savings Act of 2001, which would bring competition back into the prescription drug industry by giving consumers the ability to purchase American-made, FDA-approved prescription drugs from other countries at lower prices.

"This legislation is a commonsense approach to solving one of the most pressing problems facing our entire health care system," she says. "It's simply unconscionable that consumers are prevented from shopping around for the best possible price for their prescription medications."

Stabenow's engaging personality and genuine warmth have made her a popular figure in her home state. She comes across as a combination big sister, mother, and friend—a person one could easily talk to across a kitchen table. But beneath the friendly exterior is a fiery resolve that has been the driving force behind her twenty-five-year political career. As her opponents have learned, she is a savvy, hard-hitting fighter who refuses to back down on the issues she cares about.

This resolve has broken new ground for women and resulted in a distinguished legislative record. At the age of twenty-four, Stabenow became the youngest person ever elected to the Ingham County Board of Commissioners, and after only two years she became the board's first woman chairperson. As an accomplished state legislator, she was a leader on tax and finance issues and was the first woman ever to hold the gavel and preside over the state's House of Representatives. More than fifty Michigan public acts bear her name. As a leader on family and children's issues, she passed reforms in Michigan for the enforcement of child support and visitation that became the model for federal law in the 1980s. And she led the nation in creating the first Children's Trust Funds for the prevention of child abuse and neglect.

Stabenow's agenda this day is filled with a series of visitors from the state. "It's so important to stay connected with people," she says. "I think of Washington, D.C., as a long-distance commute from my home—not as a second home. I go back to Michigan nearly every weekend. That's where my church, my grocery store, and my dry cleaner are. I'm very aware of how easy it is to get so wrapped up in what's going on in the capital that you don't consider the impact it has on the people at the local level. That's why I've made it a priority to stay connected to Michigan."

Among her visitors is a group representing Habitat for Humanity. One of its members is a good friend from Grace United Methodist Church in Lansing—Stabenow's local church. They are here to discuss Stabenow's support for an ambitious plan to build at least one hundred houses in the coming year, two per state, under the heading "The House the United States Senate Built." Stabenow has already agreed to sponsor three houses in Michigan, but she makes a point of cautioning the group to give credit where it's due. "I'm lending my support and my visibility to help the program," she tells her visitors. "And I'll spend several hours at each house. But make it clear that the community is doing the work, not me. They need to get the credit."

Stabenow's life is grounded in family, church, and community. She defines herself through these connections. She has been active in her church, and believes that the church has a key role to play as a socially responsible institution serving as the supportive framework of the community. She learned this from her parents at an early age. Although they weren't involved in politics, their activities in church and community set an example for their children. From them, Stabenow came to appreciate grass-roots activism as the source of change.

A second-term congresswoman in 1999, Stabenow announced her intention to challenge Senator Spencer Abraham for his senate seat. Many of her fellow Democrats were concerned that the risk was too great. But Stabenow believed that the first-term senator was vulnerable. She was eager to present the people of Michigan with what she saw as a better choice—that of an advocate on the bread-and-butter issues so critical to a struggling industrial state.

Offering early encouragement was Ellen Malcolm, president of EMILY's List, the fund-raising arm for Democratic, pro-choice women candidates. "I told Debbie that I believed she could win the race," Malcolm recalls. "I knew that she was the most qualified and dynamic candidate to take on Spencer Abraham, who was anti-choice. From the moment Debbie decided to run, EMILY's List ran with her, and we never stopped for breath until Abraham conceded on November 8."

The $1.4 million raised for Stabenow through EMILY's List helped to keep her competitive with her opponent, but she could never begin to match his war chest. A full year before the election, Abraham was already airing a million dollars' worth of commercials attacking Stabenow. But the race remained a dead heat. For Stabenow was able to offer Michigan voters a clear choice. Her defining issues were those about which they cared, and they elected her by a narrow margin. Throughout the campaign, her mother; her son, Todd; her daughter, Michelle; and her brothers worked as hard for her as they had in many prior campaigns.

Stabenow's twenty-five-year-old son, Todd, cut his teeth on political campaigns, as Stabenow demonstrates by referring to two pictures set side by side on her desk. One shows four-year-old Todd sitting on his bicycle, wearing a big smile and a T-shirt that says: "Vote for my Mom." The second shows a handsome young man of twenty-five, still wearing a big grin, and still wearing a T-shirt saying, "Vote for my Mom."

As a divorced, working mother, Stabenow has demonstrated by example that women do not have to sit on the sidelines of public life. For her, the sentiment "Vote for my mom" is a rallying cry for a new generation.

DEBBIE STABENOW

I take my position as a role model very seriously. It's so important that young women see public service as an option—that they see they have a place among the decision makers. I've always made it a point to tell my daughter, Michelle, the tangible ways that women in government have expanded her own opportunities. When she played sports in school, I talked to her about Title Nine. I told her that were it not for a small group of women who fought to see that girls' sports programs were funded, she may never have had the opportunity to play on teams. What they did really mattered. I told her if not for a small group of women who decided to fight for women's health issues, it might have taken far longer to break through so many barriers in women's health care and research. It was the women of the Congress who pushed for women to be fully represented in medical research. They were told that women couldn't be part of the studies because their hormonal structures made them different. Well, that was the point! Women's hormones made them different.

As a result, my daughter as well as my son see public service as a part of life. They both see politicians as approachable. They understand the importance of voting. As soon as they turned eighteen, they registered to vote.

I took my young niece with me to the Democratic National Convention last summer, to show her that she can be a part of the process. She was so excited. She came away from the convention knowing that every opportunity is available to her.

If we're going to help our democracy thrive, we have to create good *citizens*—that is, powerful, responsible, engaged members of society. I'm proud to be a model of that wonderful possibility for young women.

It has been a long and hectic day. The Senate is preparing to break for a week, and Senator Stabenow is headed to Michigan's frigid Upper Peninsula. Snow boots and heavy coats are already packed as she goes over the plans with her scheduler. Suddenly, a matter of great importance occurs to her. She jumps up from her desk, announcing, "I have to run down to my car. I'll be right back." She hurries out of her office and returns sometime later, carrying an enormous red heart-shaped box of chocolates. Stabenow removes its cover and begins walking around the office, urging her staff to put down their phones, turn away from their computers, and indulge themselves for a couple of moments. There may be little time for traditional romantic gestures in a busy Senate office, but at least a moment of camaraderie can be shared around a choice selection of delicacies. "Happy Valentine's Day, everybody," Stabenow calls out cheerfully, then disappears back into her office to tackle the next order of business.

February 15, 2001—Russell Senate Office Building

Visits by constituents are commonplace and welcome in the offices of the Senate—an ever-present reminder that senators derive their power from the people. That is why six schoolteachers from Washington State have settled in to the cramped waiting area of Maria Cantwell's temporary office one afternoon, awaiting her return from the Senate floor. They are surrounded by stacks of files, boxes, and artifacts from their home state. A photograph of Mount Saint Helens leans against a table; a map of Washington State is spread out on the floor. Busy staff members rush in and out of the room, gliding past the teachers and stepping carefully around the clutter. The teachers don't seem to mind the wait. They chat excitedly about meeting their new senator.

When Cantwell finally does appear, striding quickly toward her office for an overdue meeting, she pauses to greet the teachers, graciously shaking their hands and thanking them for stopping by.

"We're so thrilled to have you in the Senate," one of the teachers effusively exclaims. She rustles around in her bag and pulls out a large package. "The children have made you something for your office," she says. The package is unwrapped to reveal a framed collage of a fish—another artifact to decorate the senator's office. Cantwell is touched by the gesture, and she carefully checks the back of the picture to make sure the children's names are listed so she can send thank-you notes. The teachers pull out their cameras and Cantwell poses with them, holding the children's collage and smiling.

Barely a month into her term, Maria Cantwell is thriving despite the chaos. In fact, when she is asked how she manages to find any balance in her life, she jokes, "Compared to the software industry, the Senate is a very balanced place."

In 1980, when Maria Cantwell first arrived in Washington State, it was love at first sight. She was twenty-two years old, had just graduated from Miami University of Ohio, and was ready to begin her first job, working for California senator Alan Cranston's presidential campaign. Cranston's bid for the White House was unsuccessful, but by the time the doors of his Washington State headquarters closed, Cantwell had already decided to make the Pacific Northwest her home. She was drawn to its spectacular beauty, the friendly openness of its residents, its growing reputation as a key player in the global economy, and its progressive politics. She settled in Mountlake Terrace, a lovely setting dominated by evergreens just north of Seattle with stunning views of the Cascade Mountains, sparkling lakes, and abundant parkland.

Although her first job had been in politics, Cantwell wasn't planning on a political career. Like many of the women who would eventually make their way to the United States Senate, she was drawn into public service through her involvement with a local issue. Mountlake Terrace needed a library, and Cantwell successfully led the campaign. Her neighbors were impressed, and they urged her to run for the state legislature. "I had always thought that if I ever did run for office, it would be much later in life," she says, "not when I was twenty-eight." As she would do on several occasions in her life, Cantwell decided to seize the moment. She knocked on every door in her district and won the election. She served eight years in the Washington Legislature, where she established a reputation as a consensus-builder and an untiring worker with a quiet, focused manner. She was "a work-horse, not a show horse," according to former State House Speaker Joe King. "She kept her mouth shut and her head down for a while," King recalls. "She wasn't too pushy. She just came in and quietly went to work. She was the best legislator I ever served with."

In 1992, Cantwell won a seat in Congress, representing the First District, north of Seattle. By then Washington was fast becoming the epicenter of the burgeoning computer technology industry, and Cantwell's district contained many of the world's most influential software and technology companies. In Congress, Cantwell helped defeat the "Clipper Chip," a Clinton administration proposal that raised privacy concerns. The bill would have required computer manufacturers to install a chip in their units allowing law enforcement to countermand encryption software.

After she lost a reelection bid in 1994, Cantwell plunged into another new challenge. She joined a Seattle-based software company called RealNetworks, which was involved in developing groundbreaking video and audio technology for the Internet. She became the senior vice president of the consumer and E-commerce divisions. Cantwell's time at RealNetworks was so rewarding that she was able to fund her own campaign when she decided to make a run for the Senate. A strong supporter of campaign finance reform, Cantwell used this advantage as one of the themes of her campaign. "The only special interests I'll have to answer to is you," she told voters. (One of her first acts as senator was cosponsoring the McCain-Feingold campaign finance reform bill.)

Cantwell joined the Senate race only six months before election day, setting off on a bus tour of Washington's thirty-nine counties that was dubbed the "One Washington Tour." The theme was to become a staple of her campaign. "I believe it is time for one Washington, not two," Cantwell declared when she announced her candidacy. "A Washington where hunters are not pitted against Indians, where farmers are not pitted against environmentalists, where rural families are not pitted against urban ones, and where new-economy workers are not pitted against old-economy workers."

Cantwell faced a strong primary challenger in fellow Democrat Deborah Senn, the state insurance commissioner who had twice won statewide office. After winning the primary, Cantwell faced a more formidable challenge still—defeating four-term incumbent Republican Slade Gorton, who had held public office for forty of the past forty-two years.

Cantwell's strategy was to reiterate basic Democratic issues, but to further position herself as a trade-oriented "New Democrat" with a sterling record of consensus and achievement while serving both as state legislator and congresswoman. "I've had a legislative career," Cantwell reminded voters. "You know you have to bring people together to get legislation passed." She was one of the chief architects of Washington State's 1990 Growth Management Act, a complex piece of legislation designed to limit urban sprawl and the depletion of natural resources.

By election day, polls showed a tight race, but no one would have guessed that election night would last three and a half weeks. The result was close enough to require a state-ordered recount. In addition, there was a record-breaking number of absentee ballots to be counted. The coming weeks were a dizzying seesaw of projections. First Cantwell was declared the winner. Then, a few days later, Gorton was so certain he'd won, he scheduled his victory announcement. But it was Maria Cantwell who finally prevailed, officially declaring victory on December 1. The victory was elating, but it taught Cantwell a lesson she doubts she'll ever forget: "The only day that really matters in a recount is the last day." Less than a week later, Cantwell was in Washington, D.C., joining her fellow freshmen in orientation. She hasn't paused since.

MARIA CANTWELL

When I was growing up in Indiana, I never actually thought, "I'm going to run for office someday." But our parents instilled in us a sense of obligation to others. They taught us that we had a duty to participate in the process of government. Both of them were very active in the Democratic Party. My father was a county commissioner, then a city councilman in Indianapolis. I remember him sitting at the kitchen table, talking to the people from the community who had come to tell him about their problems. It made a lasting impression on me.

My father always told me I could do anything I set out to do. He never treated me differently from my brothers, and I never felt that I was held back because I was a woman. That was of vital importance to the way I've conducted my entire life. The only real obstacles are the ones that we create in our own minds, not the ones that we imagine are there.

I'd received my degree in public policy, and I thought of myself as more of a policy person than a politician. Running for office wasn't on my mind. But after we were able to get the library funded, some of the people in my neighborhood urged me to run for the state legislature, and so I did. Suddenly, there I was, elected to the legislature at the age of twenty-eight.

When I ran for Congress in 1992, I wasn't sure I had a chance to make it. My district had been voting almost entirely Republican for about forty-two years. It was a terrific challenge, but I was elected. When I didn't get reelected in 1994, I moved on to the next challenge and did something completely different.

I think Washington State is very open to women in government. Now we're the third state to have two women in the Senate. During the Senate race, my being a woman never came up as an issue. People didn't say, "We already have one woman, Patty Murray, in the Senate. We don't need two." If anything, it might have been an advantage. Women still face barriers in getting elected to office, but I've seen a tremendous change. In 1992, when I ran for Congress, my opponent tried to make an issue of my being single. He thought the voters would rather send a "family man" to Congress. Today, those issues don't get raised because people understand that women play many different roles. During my campaign, I noticed how enthusiastic young women were about my candidacy. They identified with me. They liked the fact that I could succeed in business and in political life. I hope I can be a role model for a new generation of women. We need that energy in government.

October 16, 2000—Jefferson City, Missouri

It has been said that a single moment can change a life, and with it the course of history. That moment occurred for Jean Carnahan on the evening of October 16 when she looked up from her desk and met the eyes of a Missouri state trooper standing in the doorway. Before the interruption, she had been working on a speech to be delivered on behalf of her husband, Governor Mel Carnahan, the following day. The speech would never be delivered.

Without a word, the state trooper walked across the room and bent to one knee. Grasping Carnahan's hand in his own, he told her, in a voice choked with tears, that her husband; her eldest son, Randy; and Chris Sifford, a top aide, had died in the crash of their small plane, piloted by Randy.

Jean Carnahan remembers the week that followed as a blur. But others in Missouri's Democratic Party were trying to staunch their grief by planning for the immediate future. As the news of Mel Carnahan's death spread, the initial reaction of shock and loss tentatively began to coalesce into something more tangible— possibility. It was too late to remove Mel Carnahan's name from the ballot, and in any case, the formidable Republican senator John Ashcroft was running strongly. Before the crash, the two contenders were running neck and neck, with Governor Carnahan's polls showing he had a slight lead.

Within forty-eight hours, a whispering campaign had started to gain momentum. By the day of the memorial service, the whispers had risen to a roar. Before the service a group of ministers from the St. Louis Clergy Coalition paid a private condolence visit to Jean Carnahan. "The governor would want us to move forward with his agenda," they told her. "We must pass the baton." She was grateful for their sentiments, although she wasn't registering their full message. But after the memorial service, as the family transported the governor's casket back to their home in Rolla, Missouri, Jean Carnahan was struck by the sight of hundreds of people lining the streets cheering for her husband. Within a few short days, those cheers had turned into resolve.

A week after Mel Carnahan's death, his Senate campaign staff was back at work, preparing to launch a dramatic and daring effort. The word went out: Vote for Mel Carnahan. Buttons flooded the streets: "I'm still with Mel." Governor Roger Wilson announced that if Mel Carnahan should win, he would be in favor of appointing Jean Carnahan to the seat. Would she accept?

"I thought long and hard about it," Carnahan says. "I kept coming back to what Mel used to say when he'd get up from the table and leave the breakfast room hearth. 'Don't let the fire go out.' And I always knew he meant it, too. I thought about it again and again. I didn't want to let all the things Mel and I had spent years working together for just come to an end. And I knew that the people of Missouri wanted something to survive that plane crash as well. If my husband could no longer be there in the flesh, then I could show that his spirit was indomitable, and deserved to live on."

And so she decided to say yes. Jean Carnahan had often been praised for her strength, sensitivity, and grace under pressure. Now she would have her mettle put to the test yet again.

For the next two weeks Jean Carnahan turned her private mourning into public action on the campaign trail. As November 7 dawned, the race was too close to call. That evening, surrounded by her three surviving children, close friends, and campaign aides, Jean Carnahan turned on her television for the first time since the crash that had changed her life and settled in to await the election results.

Long after midnight, when Mel Carnahan was finally declared the winner of Missouri's Senate race, there was silence in the room. "There were no cheers," Carnahan remembers. "There were only tears."

On December 4, Governor Wilson officially appointed Jean Carnahan to a two-year term in the United States Senate. The following day, Carnahan headed for freshman orientation in Washington. When a reporter asked her if she thought she was up to the job, she didn't even blink. "A lot of people spent their lives underestimating my husband," she said with a spark of fire in her eyes. "I suggest that they not start doing that with me."

In 1993, on Jean Carnahan's first morning in the Missouri governor's mansion, she went down to the kitchen to get a cup of coffee. While she was searching for a cup, one of the prison inmates who worked at the mansion asked her what she was looking for. She replied, "Oh, I'm just looking for a Styrofoam cup for my coffee."

He reacted with horror. "Mrs. Carnahan, the First Lady doesn't drink coffee out of a Styrofoam cup."

She replied, "Oh, please forgive me, I didn't know that!"

This quality of unpretentious good humor will certainly come in handy as Carnahan makes the transition to the United States Senate. Her temporary quarters on the fourth floor of the

Russell Senate Office Building are crammed with unpacked boxes, and staff members are squeezed into tiny cubicles. Carnahan's private office is a claustrophobic corner room barely large enough to hold a desk and a bookcase, but she seems relaxed, and there's a twinkle in her eye as she talks about the frenzy of her new life. She has just moved into a two-bedroom rental apartment in Washington, and is amused to find herself, at age sixty-seven, bargain shopping for furniture and household necessities. "Suddenly last night I realized there was no iron in my apartment," she says. "Fortunately, Maria Cantwell told me she has three irons—she bought one herself, then received two as gifts. So she's bringing one today." She laughs as she contemplates the sight of senators exchanging irons, toasters, and teapots on the Senate floor. Perhaps the citizenry would take comfort in knowing that senators really do share the down-to-earth daily concerns of their constituents. It's hard to sit on a pedestal when you have to run out after a vote to shop for toilet paper, dish soap, and lightbulbs.

As with many of the women senators, much of Carnahan's appeal comes from the fact that she never set out to be a professional politician. That was her husband's dream. She met Mel Carnahan, the son of a congressman from Missouri, at a church group when they were both fifteen. He already knew what he wanted. On their second date, he declared his intentions with great surety, saying, "I'm going to marry you and I'm going to run for office." Five years later they were married, and five years after that he won his first election, to the post of municipal judge in Rolla, Missouri. During their forty-five-year marriage, Jean Carnahan stayed in the background, raising their four children, writing her husband's speeches, and serving as his de facto campaign manager during twenty political campaigns. When Mel Carnahan was elected governor of Missouri, she saw that she had a new opportunity to make an important contribution of her own. As First Lady, Jean Carnahan was highly visible, lobbying tirelessly for her husband's programs. She was instrumental in the promotion of his Outstanding Schools Act, designed to empower children to learn through the advantages of smaller class size and new technology. She was involved in efforts to make sure every child in Missouri received immunizations. She opened the mansion to the people, hosting events and welcoming large groups into her home. She also wrote three books—*If Walls Could Talk: The Story of Missouri's First Families; Christmas at the Mansion: Its Memories and Menus*; and *Will You Say a Few Words?*, a book of her speeches that demonstrates her wit, warmth, and wisdom, as well as her appreciation of the power of brevity.

JEAN CARNAHAN

When we first came to the governor's mansion, I knew I had a choice. I could rearrange the furniture and host teas, or I could make some kind of lasting impact. I was very aware of being a part of history and the obligation that comes with that. I wanted to share that history with the people of Missouri. Since I had always loved to write, my first book, *If Walls Could Talk*, was a wonderful experience. Every time my researcher came up with a new detail we'd been unaware of, we'd get so excited. I felt as if we were giving the people of the state back their story.

As I was exploring the history of the mansion, I learned that there had once been a small fountain on the lawn that had long ago been removed. I found a picture of the fountain from 1910, and saw that it depicted Governor Hadley's three children playing in the basin. That gave me the idea for erecting a fountain that incorporated the figures of three children playing in the falling water. I raised the money and commissioned a fine artist, Jamie Anderson, to create the Missouri Children's Fountain. I wanted the three figures to be lasting reminders of our commitment to the health, opportunity, and environment of our children. One figure represents Carrie Crittendon, the nine-year-old daughter of one of our nineteenth-century governors, Thomas Crittendon. Little Carrie died of diphtheria, and was the only child ever to die at the mansion. She was to remind us of the health needs of children. Another figure depicts an African American boy reaching into the water. He came to mind when I read a reference that First Lady Agnes Hadley made to the "little colored boy who stays in the barn from time to time." I couldn't get that little boy out of my mind. Why was he there? Had he run away, been abandoned or orphaned? The African American boy is there as a reminder that no child should be left out. The third child in the fountain represents the children of today. He is surrounded by water, birds, fish, and oak leaves, and he reminds us that we must preserve our environment for future generations.

The fountain is an example of how art can speak to us, sometimes more powerfully than words, to keep our concerns and commitments at the forefront. I'm very proud of that fountain.

Now here I am, beginning an entirely unexpected new stage in my life. It wasn't what I'd expected. I had hoped to be here as a senator's spouse, not a senator. But it was never an option for me to say no to this. My husband and I had worked so hard for so long on the issues we believed in. I couldn't walk away. For now, I realize I am still in his shadow. When I speak to groups, they want to hear me talk about Mel, and what brought me here. And that's fine. I understand. It helps people feel a connection. I think it's a good thing. It's cathartic. I can't tell you how many times people have come up to me and started sharing the stories of their own losses. They can talk to me because they know I've been there too. We have a real bond.

I have a lot to learn and a lot to work out in my own mind about what I will accomplish here. Right now, every aspect of my life is in transition. But I know that at some point I'm going to have to step back and do some things on my own. I want to be judged by my own efforts, my own commitment to the issues, not my husband's.

I take comfort in something Eleanor Roosevelt once said— "Women have one advantage over men. Throughout history they have been forced to make adjustments. The result is that, in most cases, it is less difficult for a woman to adapt to new situations than it is for a man." Right now, I'm hoping that's true.

In the history of our nation, only 1,864 citizens have served in the United States Senate. Thirty-one have been women. That number will most certainly rise, but the route to equity may continue to be circuitous for the present generation. In 2002, two of the Senate's women—Maine Republican Susan Collins and

Louisiana Democrat Mary Landrieu—will face reelection in what already promise to be very tough races. Perhaps it is a sign of how effective Collins and Landrieu have been that their opponents feel the need to start working especially early. Jean Carnahan's seat will also be open, and she has vowed to run for election only if she knows she's made a difference. It is likely that more women will emerge, from both parties, to enter races across the country.

The era of inclusion has arrived. As increasing numbers of women gain political power, they have begun to dispel the myths and biases that have held women back. Today, membership in the "world's most exclusive club" is open to the women of America.

APPENDIX A

The Women of the United States Senate: 1922–2001

An asterisk denotes those who have been elected to one or more full terms.

1. **Rebecca Latimer Felton** (D) Georgia
   11/21/22–11/22/22
   Appointed to fill the term of deceased senator Thomas E. Watson.
2. **Hattie Wyatt Caraway** (D) Arkansas*
   11/13/31–01/02/45
   Appointed to fill the term of her deceased husband, Thaddeus H. Caraway, then was reelected to two full terms.
3. **Rose McConnell Long** (D) Louisiana
   01/31/36–01/02/37
   Received an interim appointment to the seat of her assassinated husband, Huey P. Long.
4. **Dixie Bibb Graves** (D) Alabama
   08/20/37–01/10/38
   Received an interim appointment to the seat of Senator Hugo Black, who resigned to become a U.S. Supreme Court justice.

5. **Gladys Pyle** (R) South Dakota
   11/09/38–01/03/39
   Elected to fill the term of deceased senator Peter Norbeck.
   Previous Service: South Dakota House of Representatives, 1923–27
   South Dakota secretary of state, 1927–31

6. **Vera Cahalan Bushfield** (R) South Dakota
   10/06/48–12/26/48
   Appointed to fill the term of her deceased husband, Harlan J. Bushfield.

7. **Margaret Chase Smith** (R) Maine*
   01/03/49–01/03/73
   Elected to four terms in the U.S. Senate.
   Previous Service: U.S. House of Representatives, 1941–49

8. **Eva Kelly Bowring** (R) Nebraska
   04/26/54–11/07/54
   Received an interim appointment to the seat of deceased senator Dwight Griswold.

9. **Hazel Hempel Abel** (R) Nebraska
   11/08/54–12/31/54
   Elected to fill the final months of Dwight Griswold's term.

10. **Maurine Brown Neuberger** (D) Oregon*
    11/08/60–01/02/67
    Elected simultaneously to the remainder of her deceased husband Richard Neuberger's term and to a six-year term.
    Previous Service: Oregon House of Representatives, 1951–55

11. **Elaine Schwartzenburg Edwards** (D) Louisiana
    08/07/72–11/13/72
    Received an interim appointment from her husband, Governor Edwin Edwards, to fill the term of deceased senator Allen Ellender.

12. **Muriel Buck Humphrey** (D) Minnesota
    02/06/78–11/07/78
    Appointed to fill the term of her deceased husband, Hubert H. Humphrey.

13. **Maryon Pittman Allen** (D) Alabama
    06/08/78–11/07/78
    Appointed to fill the term of her deceased husband, James B. Allen.

14. **Nancy Landon Kassebaum** (R) Kansas*
    12/23/78–01/06/97
    Elected to three terms in the U.S. Senate.
    Previous Service: Maize, Kansas, School Board, 1972–75

15. **Paula Fickes Hawkins** (R) Florida*
    01/01/81–01/03/87
    Elected to one term in the U.S. Senate.
    Previous Service: Florida Public Service Commission, 1973–79

16. **Barbara Mikulski** (D) Maryland*
    01/03/87–
    Elected to, and currently serving, her third term in the U.S. Senate.
    Previous Service: Baltimore, Maryland, City Council, 1971–76
    U.S. House of Representatives, 1977–86

17. **Jocelyn Birch Burdick** (D) North Dakota
    09/16/92–12/14/92
    Received an interim appointment to fill the term of her deceased husband, Quentin Burdick.

18. **Dianne Feinstein** (D) California*
    11/10/92–
    Elected in 1992 to the two years remaining in Pete Wilson's term. Wilson resigned in 1990 to become governor, and appointed John Seymour to hold the seat until the 1992 special election. Reelected to full terms in 1994 and 2000.
    Previous Service: Member/President, San Francisco, California, Board of Supervisors, 1970–78
    Mayor, San Francisco, California, 1978–88

19. **Barbara Boxer** (D) California*
    01/05/93–
    Elected to, and currently serving, her second term in the U.S. Senate.
    Previous Service: Marin County, California, Board of Supervisors, 1976–82
    U.S. House of Representatives, 1983–92

20. **Carol Moseley-Braun** (D) Illinois*
    01/05/93–01/04/99
    Elected to one term in the U.S. Senate.
    Previous Service: Illinois House of Representatives, 1978–88
    Cook County, Illinois, recorder of deeds, 1989–92

21. **Patty Murray** (D) Washington*
    01/05/93–
    Elected to, and currently serving, her second term in the U.S. Senate.
    Previous Service: Shoreline, Washington, School Board, 1985–89
    Washington State Senate, 1988–92

22. **Kay Bailey Hutchison** (R) Texas*
    06/14/93–
    Won a 1993 special election to fill the term of Lloyd Bentsen, who resigned to accept a cabinet position as secretary of the treasury. Reelected to full terms in 1994 and 2000.
    Previous Service: Texas House of Representatives, 1972–76
    Texas state treasurer, 1990–93

23. **Olympia J. Snowe** (R) Maine*
    01/04/95–
    Elected to, and currently serving, her second term in the U.S. Senate.
    Previous Service: Maine House of Representatives, 1973–77
    Maine State Senate, 1977–79
    U.S. House of Representatives, 1979–95

24. **Sheila Sloan Frahm** (R) Kansas
    06/11/96–11/27/96
    Received an interim appointment after Robert Dole resigned to accept the Republican presidential nomination.
    Previous Service: Colby, Kansas, School Board, 1981–85
    Kansas State Board of Education, 1985–89
    Kansas State Senate, 1989–95
    Lieutenant governor, 1995–96

25. **Susan Collins** (R) Maine*
    01/07/97–
    Elected to, and currently serving, her first term in the U.S. Senate.

26. **Mary Landrieu** (D) Louisiana*
    01/07/97–
    Elected to, and currently serving, her first term in the U.S. Senate.
    Previous Service: Louisiana House of Representatives, 1979–87
    Louisiana state treasurer, 1987–95

27. **Blanche Lambert Lincoln** (D) Arkansas*
    01/04/99–
    Elected to, and currently serving, her first term in the U.S. Senate.
    Previous Service: U.S. House of Representatives, 1993–97

28. **Debbie Stabenow** (D) Michigan*
    01/03/01–
    Elected to, and currently serving, her first term in the U.S. Senate.
    Previous Service: Ingham, Michigan, county commissioner, 1975–78
    Michigan House of Representatives, 1979–90
    Michigan State Senate, 1991–94
    U.S. Congress, 1997–2000

29. **Maria Cantwell** (D) Washington*
    01/03/01–
    Elected to, and currently serving, her first term in the U.S. Senate.
    Previous Service: Washington State House of Representatives, 1987–93 U.S. House of Representatives, 1993–94

30. **Hillary Rodham Clinton** (D) New York*
    01/03/01—
    Elected to, and currently serving, her first term in the U.S. Senate.

31. **Jean Carnahan** (D) Missouri
    01/03/01—
    Appointed to an interim term of two years to the seat won by her deceased husband, Governor Mel Carnahan. Faces reelection in 2002.

APPENDIX B

Women's Work 2001

The thirteen women currently serving in the United States Senate have been responsible for authoring hundreds of pieces of legislation—taking direct action on some of America's foremost concerns. Their committee assignments and legislative achievements cover a wide range of issues. The following is contact and committee information for each senator. Their web pages offer detailed information about legislation they have authored and new laws that have been enacted based on their efforts.

If you would like to learn more about Senate business, history, and members, and be able to read full descriptions of the business of Senate committees and the legislative actions of Congress, go to the Senate website at http://www.senate.gov.

SENATOR BARBARA BOXER
Democrat, California
112 Hart Senate Office Building
Washington, D.C. 20510
Phone: (202) 224-3553
Fax: (202) 228-1338
E-mail: senator@boxer.senate.gov
Web: http://boxer.senate.gov

SENATE LEADERSHIP
Chief deputy for strategic outreach, Democratic Conference

COMMITTEES
Committee on Commerce, Science, and Transportation
   Subcommittees:
   Communications
   Oceans and Fisheries
   Consumer Affairs, Foreign Commerce, and Tourism
   Surface Transportation and Merchant Marine
Committee on Environment and Public Works
   Subcommittees:
   Ranking Minority Member: Superfund, Waste Control, and Risk
      Assessment
   Transportation and Infrastructure
Committee on Foreign Relations
   Subcommittees:
   Ranking Minority Member: International Operations and Terrorism
   African Affairs
   Near Eastern and South Asian Affairs

SENATOR MARIA CANTWELL
Democrat, Washington
717 Hart Senate Office Building
Washington, D.C. 20510
Phone: (202) 224-3441
Fax: (202) 228-0514
E-mail: maria@cantwell.senate.gov
Web: http://cantwell.senate.gov

COMMITTEES

Committee on Energy and Natural Resources
　Subcommittees:
　Energy Research, Development, Production, and Regulation
　Forests and Public Land Management
　Water and Power
Committee on the Judiciary
Committee on Small Business

SENATOR JEAN CARNAHAN
Democrat, Missouri
517 Hart Senate Office Building
Washington, D.C. 20510
Phone: 202-224-6154
E-mail: senator_carnahan@carnahan.senate.gov
Web: http://carnahan.senate.gov

COMMITTEES

Committee on Armed Services
　Subcommittees:
　Sea Power
　Airland
　Personnel

Committee on Commerce, Science, and Transportation
   Subcommittees:
   Aviation
   Consumer Affairs, Foreign Commerce, and Tourism
   Science, Technology, and Space
   Surface Transportation and Merchant Marine
Committee on Governmental Affairs
Special Committee on Aging

SENATOR HILLARY RODHAM CLINTON
Democrat, New York
476 Russell Senate Office Building
Washington, D.C. 20510
Phone: (202) 224-4451
Fax: (202) 228-0282
E-mail: senator@clinton.senate.gov
Web: http://clinton.sentate.gov

COMMITTEES
Commission on Security and Cooperation in Europe
Committee on the Budget
Committee on Environment and Public Works
   Subcommittees:
   Clean Air, Wetlands, Private Property, and Nuclear Safety
   Fisheries, Wildlife, and Water
   Superfund, Waste Control, and Risk Assessment
Committee on Health, Education, Labor, and Pensions
   Subcommittees:
   Aging
   Public Health

SENATOR SUSAN COLLINS
Republican, Maine
172 Russell Senate Office Building
Washington, D.C. 20510
Phone: (202) 224-2523
Fax: (202) 224-2693
E-mail: senator@collins.senate.gov
Web: http://collins.senate.gov

COMMITTEES

Committee on Armed Services
   Subcommittees:
   Seapower
   Emerging Threats and Capabilities
   Personnel
Committee on Governmental Affairs
   Subcommittees:
   Chairman: Permanent Subcommittee on Investigations
   International Security, Proliferation, and Federal Services
Committee on Health, Education, Labor, and Pensions
   Subcommittees:
   Children and Families
   Public Health
Special Committee on Aging

SENATOR DIANNE FEINSTEIN
Democrat, California
331 Hart Senate Building
Washington, D.C. 20510
Phone: (202) 224-3841
Fax: (202) 228-3954
E-mail: senator@feinstein.senate.gov
Web: http://feinstein.senate.gov

COMMITTEES

Committee on Appropriations

   Subcommittees:

   Ranking Minority Member: Military Construction

   Defense

   Agriculture, Rural Development, and Related Agencies

   Interior

   Energy and Water Development

Committee on Energy and Natural Resources

   Water and Power

   Energy Research and Development

   Forest and Forest Health

Committee on the Judiciary

   Subcommittees:

   Ranking Minority Member: Technology, Terrorism, and Government Information

   Immigration

   Youth Violence

Committee on Rules and Administration

Select Committee on Intelligence

Joint Committee on Printing

United States Senate Caucus on International Narcotics Control

United States Senate Cancer Coalition

SENATOR KAY BAILEY HUTCHISON

REPUBLICAN, TEXAS

284 Russell Senate Office Building

Washington, D.C. 20510

Phone: (202) 224-5922

Fax: (202) 224-0776

E-mail: senator@hutchison.senate.gov

Web: http://hutchison.senate.gov

SENATE LEADERSHIP
Vice chairman, Senate Republican Conference

COMMITTEES
Committee on Appropriations
    Subcommittees:
    Chairman: Military Construction
    Defense
    District of Columbia
    VA, HUD, and Independent Agencies
    Labor, Health and Human Services, and Education
    Military Construction
    Commerce, Justice, State, and the Judiciary
Committee on Commerce, Science, and Transportation
    Subcommittees:
    Chairman: Aviation
    Surface Transportation and Merchant Marine
    Communications
    Science, Technology, and Space
    Oceans and Fisheries
Committee on Rules and Administration

SENATOR MARY LANDRIEU
Democrat, Louisiana
724 Hart Senate Office Building
Washington, D.C. 20510
Phone: (202) 224-5824
Fax: (202) 224-9735
E-mail: http://landrieu.senate.gov/webform.html
Web: http://landrieu.senate.gov

COMMITTEES

Committee on Appropriations
  Subcommittees:
  Ranking Minority Member: District of Columbia
  Foreign Operations
  Military Construction
  Labor, Health and Human Services, and Education
  Treasury and General Government

Committee on Armed Services
  Subcommittees:
  Ranking Minority Member: Emerging Threats and Capabilities
  Seapower
  Readiness and Management Support

Committee on Energy and Natural Resources
  Subcommittees:
  Energy Research, Development, Production, and Regulation
  Forests and Public Land Management
  National Parks, Historic Preservation, and Recreation

Committee on Small Business

SENATOR BLANCHE LINCOLN
Democrat, Arkansas
355 Dirksen Senate Office Building
Washington, D.C. 20510
Phone: (202) 224-4843
Fax: (202) 228-1371
E-mail: blanche_lincoln@lincoln.senate.gov
Web: http://lincoln.senate.gov

COMMITTEES
Committee on Agriculture, Nutrition, and Forestry

Subcommittees:

Ranking Minority Member: Forestry, Conservation, and Rural Revitalization

Production and Price Competitiveness

Committee on Finance

Subcommittees:

Health Care

International Trade

Taxation and IRS Oversight

Select Committee for Ethics

Special Committee on Aging

SENATOR BARBARA MIKULSKI

Democrat, Maryland

709 Hart Senate Office Building

Washington, D.C. 20510

Phone: (202) 224-4654

Fax: (202) 224-8858

E-mail: senator@mikulski.senate.gov

Web: http://mikulski.senate.gov

SENATE LEADERSHIP

Secretary of Democratic Conference

COMMITTEES

Committee on Appropriations

Subcommittees:

Ranking Minority Member: VA, HUD, and Independent Agencies

Foreign Operations

Commerce, Justice, State, and the Judiciary

Transportation

Treasury and General Government

Committee on Health, Education, Labor, and Pensions
    Subcommittees:
    Ranking Minority Member: Aging
    Public Health

SENATOR PATTY MURRAY
Democrat, Washington
173 Russell Senate Office Building
Washington, D.C. 20510
Phone: (202) 224-2621
Fax: (202) 224-0238
E-mail: senator_murray@murray.senate.gov
Web: http://murray.senate.gov

SENATE LEADERSHIP
Chair, Democratic Senatorial Campaign Committee

COMMITTEES
Committee on Appropriations
    Subcommittees:
    Ranking Minority Member: Transportation
    Labor, Health and Human Services, and Education
    Commerce, Justice, State and the Judiciary
    Interior
    Energy and Water Development
Committee on the Budget
Committee on Health, Education, Labor, and Pensions
    Subcommittees:
    Children and Families
    Aging
Committee on Veterans' Affairs

SENATOR OLYMPIA J. SNOWE
Republican, Maine
154 Russell Senate Office Building
Washington, D.C. 20510
Phone: (202) 224-5344
Fax: (202) 224-1946
E-mail: olympia@snowe.senate.gov
Web: http://snowe.senate.gov

SENATE LEADERSHIP
Counsel to the assistant majority leader

COMMITTEES
Commission on Security and Cooperation in Europe
Committee on the Budget
Committee on Commerce, Science, and Transportation
   Subcommittees:
   Chairman: Oceans and Fisheries
   Aviation
   Surface Transportation and Merchant Marine
   Communications
Committee on Finance
   Subcommittees:
   Chairman: Health Care
   International Trade
   Taxation and IRS Oversight
Committee on Small Business

SENATOR DEBBIE STABENOW
Democrat, Michigan
702 Hart Senate Office Building
Washington, D.C. 20510
Phone: (202) 224-4822
E-mail: senator@stabenow.senate.gov
Web: http://stabenow.senate.gov

COMMITTEES

Committee on Agriculture, Nutrition, and Forestry
   Subcommittees:
   Forestry Conservation and Rural Revitalization
   Research, Nutrition, and General Legislation
Committee on Banking, Housing, and Urban Affairs
   Subcommittees:
   Securities and Investment
   Financial Institutions
   Housing and Transportation
Committee on the Budget
Special Committee on Aging

APPENDIX C

Biographies

BARBARA BOXER

A forceful advocate for families, children, consumers, the environment, and her state of California, Barbara Boxer became a United States senator in January 1993 and was elected to a second, six-year term in 1998.

Boxer came to the Senate from the House of Representatives where she served for ten years, making her mark as a champion of human rights, environmental protection, military procurement reform, and a woman's right to choose. Boxer disclosed the famous $7,600 coffee pot and passed over a dozen procurement reforms, one of which, alone, has saved taxpayers over $2.6 billion.

Throughout her public-service career, Boxer has won numerous awards for her efforts to create a cleaner, healthier environment. As a senator, she successfully amended the Safe Drinking Water Act to ensure that standards for drinking water are set to protect the most vulnerable Americans, including children, pregnant women, and the elderly. She has introduced legislation to restore our nation's wetlands and remove the threat of offshore drilling along California's coast and is the author of the Children's Environmental Protection Act, which would require environmental standards to be set at levels that protect children.

Boxer has drawn lessons from her own experience in public education—from kindergarten through college—which has made her a champion of maintaining and improving our public schools. Her Early Education Bill would give children, especially those from low-income families, a jump-start on learning by providing grants to school districts to offer classes a year before kindergarten. Her "Computers in Classrooms" law encourages the donation of computers and software to schools and helps give students the tools they need to get a top-quality education.

Senator Boxer is also committed to making our schools and neighborhoods safe for families. Since first introducing her After School Education and Anti-Crime Act in 1997, she has worked to increase funding for after-school programs from $1 million in fiscal year 1997 to $846 million in fiscal year 2001.

Committed to eliminating violence in America, Senator Boxer joined colleagues to pass the landmark 1994 Crime Bill, which led to the lowest crime rate in twenty-five years. She voted for 100,000 new police on the beat to bolster community policing and assist with anti-gang programs. To curb gun violence, the number-one killer of California's young people, Senator Boxer has worked for commonsense gun-control measures including legislation to get junk guns off our streets. Her legislation to prevent the criminal use of personal information obtained through motor vehicle records was signed into law and upheld by the U.S. Supreme Court.

She also authored the Violence Against Women Act while serving in the House and helped steer it successfully through the Senate; it, also, is now law.

A strong advocate of medical research, Senator Boxer is a leader in the drive to double funding for the National Institute of Health. An early leader in the fight against AIDS in this country, Boxer, more recently, authored bipartisan legislation to accelerate America's leadership in the international fight against HIV/AIDS and tuberculosis. These bills are now law, and the efforts have been fully funded. One of the first in Congress to recognize HMO abuses, she authored a Patients' Bill of Rights in 1997 and continues to fight for these much needed protections. She has worked to draw attention and resources to women's health issues, such as breast cancer and cardiovascular diseases. She supported expanding health coverage for children and has called attention to the high incidence of prostate cancer and the need for resources to fight this deadly disease.

The Senate's leading advocate of a woman's right to choose, Senator Boxer authored the Family Planning and Choice Protection Act and helped lead the floor fight for passage of the Freedom of Access to Clinic Entrances Act.

A champion of senior citizens, Senator Boxer has worked to preserve the safety net for older Americans. She introduced a series of bills to protect the pensions of working Americans, including the 401(k) Pension Protection Act of 1997, which was signed into law as part of the 1998 budget agreement. She continues to work to ensure solvency for Medicaid and Medicare.

Senator Boxer has been honored for her leadership in Congress by the Consumer Federation of America, the Coalition to Stop Government Waste, Planned Parenthood, the League of Conservation Voters, the Sierra Club, the Center for Environmental Education, and the American Association of University Women. She has been hailed as a champion of human rights by the Anti-Defamation League, the

Human Rights Campaign Fund, and the Leadership Conference on Civil Rights.

Senator Boxer was elected to the House of Representatives in 1982 following six years on the Marin County Board of Supervisors, where she was elected the first female president of the board. Earlier this year, Senator Boxer became the first Democratic woman to serve on the Senate Commerce Committee. She also serves on the Senate Environment and Public Works Committee, as well as the Foreign Relations Committee. She acts as Chief Deputy for Strategic Outreach for the Senate Democratic Leadership and serves as the Western Regional Democratic Whip. She is also a member of the Senate's Hispanic Caucus.

A first generation American on her mother's side, Boxer was born and raised in Brooklyn, New York, where she received her B.A. in economics from Brooklyn College. Senator Boxer and her husband of thirty-nine years live in Greenbrae, California. They have two adult children and one grandchild.

MARIA CANTWELL

Maria Cantwell was sworn in as United States senator from the state of Washington on January 3, 2001. In her first day as a U.S. senator, Cantwell offered a tribute to the former senator, Scoop Jackson, and pledged to carry on his legacy of dedicated service. Senator Max Cleland of Georgia graciously gave Cantwell his Senate desk—the desk formerly occupied by Jackson during his thirty-one years in the Senate.

Maria Cantwell was born in 1958, the second of five children, and was raised in a modest home in a working-class Irish neighborhood on the south side of Indianapolis. Her love of public service and community involvement was passed down to her by her parents and her grandparents at an early age. Her father, Paul, a construction worker by trade, served as county commissioner, city councilman, and state legislator, and was chief of staff for U.S. senator Andrew Jacob. Her mother, Rose, worked as an administrative assistant and was also active in public affairs.

Cantwell attended high school in Indianapolis, and in 1980, with the help of student loans and odd jobs, received a bachelor's degree in public policy from Miami University of Ohio. In doing so, she became the first in her family to graduate from college. After college, Cantwell took a job in Washington State. She fell in love with the Pacific Northwest and has lived there ever since.

After settling in Mountlake Terrace, a community north of Seattle, Cantwell successfully organized a coalition to build a new library there. Soon she decided to run for state office; by knocking on every door in her district, she went to Olympia at the age of twenty-eight, one of the youngest women ever elected to the state legislature.

Cantwell rapidly established a reputation as someone who could bring people together and make things happen. She's remembered by many as the architect of the state's Growth Management Act, which she shepherded through a marathon sixty-five-day session. This and other accomplishments earned her a high degree of respect among her peers.

In 1992, Cantwell was elected to the U.S. Congress in the First District north of Seattle. In Congress, she supported such landmark legislation as the Family and Medical Leave Act and the 1993 deficit reduction plan. She was known as a steadfast supporter for environmental protection and defender of a woman's right to choose.

Cantwell's district contained many of the world's most influential software and technology firms, and she applied herself to learning the issues and standing up for this vital new sector of our economy. She is well regarded in Internet circles for fighting against archaic export restrictions on software encryption products and for helping to defeat the infamous Clipper Chip proposal.

Having immersed herself in high-tech issues while in Congress, Cantwell joined a software start-up in 1995 and helped the business grow to create a thousand jobs in Washington State.

In November 2000, Cantwell was elected to the U.S. Senate, promising to fight for reform and help expand opportunity for all of Washington State.

JEAN CARNAHAN

Before taking Harry Truman's seat in the Senate, Jean Carnahan devoted her two terms as First Lady to improving the lives of Missouri's children and to bringing a new warmth and hospitality to the governor's mansion. She has been a wife, mother, grandmother, author, children's advocate, and public speaker. When her husband, Governor Mel Carnahan, ran for the U.S. Senate she was actively involved in his campaign, just as she had been in the previous nineteen campaigns he had waged during his lifetime of public service.

However, on October 16, 2000, her life moved dramatically to the forefront. Just three weeks before the November 2000 election, Mel Carnahan was killed in a plane crash along with their oldest son, Randy, and long-time aide Chris Sifford.

With her husband's name still on the ballot, Carnahan agreed to serve in his place should he be elected. On Election Day the people of Missouri showed their faith in Mel Carnahan's ideals and their confidence in his widow to continue his work on behalf of Missouri's families, giving Carnahan a 48,000-vote plurality. She took the oath of office on January 3, 2001, after having been appointed by Governor Roger Wilson.

Even before her appointment, Carnahan had shown how strongly she felt about Missouri's families and their impact on the future of the state and nation. As First Lady, she advocated for childhood immunization and created an annual arts festival for children. She was the cofounder of Children in the Workplace, a project to develop employer-supported, on-site day-care centers for working families. She frequently spoke on behalf of victims of domestic violence and for those struggling with cancer, osteoporosis, and mental health and drug problems. She raised funds for the Rape and Abuse Crisis Center and helped build homes for Habitat for Humanity.

In 1998, after five years of research and writing, Mrs. Carnahan completed her first book, *If Walls Could Talk*, a 440-page history of Missouri's First Families and the challenges they faced in public service. The following year, she published *Christmas at the Mansion: Its Memories and Menus*. In July 1999, her speech on women of achievement was selected for national publication in *Vital Speeches of the Day*. The following year, a collection of her speeches was published in paperback form under the title *Will You Say a Few Words?* In recognition of her work, Carnahan has received the Robert C. Goshorn Award for public service and the state's Martin Luther King Jr. special achievement award. In 1995, she received the Child Advocate of the Year Award from Boys' and Girls' Town of Missouri. She was named 1997 Citizen of the Year by the March of Dimes, and the 1999 Woman of the Year by the St. Louis Zonta Clubs International. She served on the board of William Woods University and helped create the Missouri Center for the Book to recognize authors from around the state.

Jean Carnahan graduated from George Washington University with a degree in business and public administration. She has three surviving children and two grandsons. Her son Russ is an attorney and Missouri legislator; Robin and Tom are also attorneys. In her new role as U.S. senator, Carnahan views herself as a centrist seeking common-sense solutions and as an advocate for Missouri jobs, schools, and families.

HILLARY RODHAM CLINTON

Hillary Rodham Clinton was elected United States senator from New York on November 7, 2000. She is the first First Lady elected to the United States Senate and the first woman elected statewide in New York.

Senator Clinton has been an advocate for children and families for more than thirty years. To build a better future for working families, she supported policies to expand the economy, raise the minimum wage and the earned income tax credit, increase tax deductions for children, and make credit more available, including microcredit loans for women entrepreneurs. Senator Clinton is dedicated to bringing jobs to upstate New York. Through tax credits for small businesses, investments in telecommunications infrastructure, technology extension programs, skills training, the restructuring of utilities, and the lowering of airfares to increase regional accessibility, she proposes to enable the economy in upstate and other regions of New York to flourish and to stem the out-migration of young New Yorkers and their families.

Appointed by President Bill Clinton in 1993 to chair the Task Force on National Health Care, Clinton worked for months meeting with families and health care professionals. The efforts of her and her task force culminated in the Health Security Act of 1994. Disappointed that the task force was unable to make more progress, she has said that the experience introduced her to the "school of smaller steps," adding that "we must continue to make progress. It's still important that we increase access to quality health care for working families." As First Lady, she led the fight to pass the Children's Health Insurance Program, providing health insurance for millions of working families. She worked to increase funding for breast cancer research and treatment for breast cancer, prostate and colon cancer, osteoporosis, and juvenile diabetes. She worked to pass strong anti-crime measures, including the Brady Bill and the Assault Weapons Ban. Clinton is a strong supporter of the HMO Patient's Bill of Rights and of action to strengthen Medicare and to include prescription drug benefits. She advocates expanding federal funding for childhood vaccinations and diseases such as asthma and epilepsy. Health care was the topic of Clinton's maiden speech in the Senate chamber.

Senator Clinton is recognized around the world as an advocate for democracy, religious tolerance, and human rights, and as a champion for women and girls, emphasizing access to education, economic opportunity, family planning, and a woman's right to choose. With her husband, former president Bill Clinton, she has worked for peace in Northern Ireland, the Balkans, and the Middle East. Her Vital Voices program has brought women together in Asia, Africa, Latin America, and Europe to encourage their increased participation in economic and political decisions.

Born in Chicago on October 26, 1947, Clinton is the daughter of Dorothy Rodham and the late Hugh Rodham. She grew up in Park Ridge, Illinois, where she attended public school. She then attended Wellesley College. A 1973 graduate of Yale Law School, Clinton was named one of the *National Law Journal*'s 100 Most Influential Lawyers in America in both 1988 and 1991. She was appointed chair of the Legal Services Corporation by President Jimmy Carter in 1977, and served as chair of the American Bar Association Committee on Women in the Profession in 1987. From 1986 to 1989, she chaired the board of the Children's Defense Fund. In 1997, she wrote the bestselling book *It Takes a Village: And Other Lessons Children Teach Us*. She contributed nearly $1 million of the author proceeds to charities dedicated to children and families. Proceeds from her book *Dear Socks, Dear Buddy: Kids' Letters to the First Pets* were given to the National Park Foundation. Her latest book, *An Invitation to the White House*, an immediate bestseller, won critical praise as a tribute to the historic home of the nation's presidents and the families who have lived there. The White House Historical Association will receive the author proceeds from that project.

SUSAN M. COLLINS

Susan M. Collins was elected to represent Maine in the United States Senate in 1996. She is the fifteenth woman in history to be elected to the Senate and holds the leadership post of deputy whip.

Senator Collins serves on the Committee on Health, Education, Labor, and Pensions, the Special Committee on Aging, and the Committee on Governmental Affairs, where she is the first freshman to chair the Permanent Subcommittee on Investigations (PSI). As chairman, Collins has focused on such consumer issues as Internet fraud, securities scams, deceptive mailings, Medicare fraud, food safety, and telephone billing fraud.

In addition to her committee assignments, Senator Collins has been appointed by the majority leader to serve on special task forces on Social Security, education, and health care policy, all issues of high priority to her.

Born December 7, 1952, Senator Collins was raised in Caribou, a small city in northern Maine for which both of her parents served as mayor. Her family runs a fifth-generation lumber business founded by her ancestors in 1844 and operated today by two of her brothers.

A 1975 magna cum laude graduate of St. Lawrence University in Canton, New York, where she was elected to Phi Beta Kappa, Senator Collins worked for former Maine senator William Cohen for twelve years, including six years as staff director of the Senate Subcommittee on Oversight of Government Management. In 1987, she joined the cabinet of Maine governor John R. McKernan Jr. as commissioner of professional and financial regulation, a position she held for five years. She then served as New England administrator of the Small Business Administration from 1992 to 1993.

In 1993, Senator Collins ran her first campaign for public office and became the first woman in Maine history to receive a major party nomination for governor after winning an eight-way Republican primary in June of 1994. She lost in the general election in the fall.

In December of 1994, Senator Collins became the founding executive director of the Center for Family Business at Husson College in Bangor, Maine, a position she held until she resigned in 1996 to run for the Senate seat being vacated by Senator Cohen. She won both a contested Republican primary and a four-way general election later that year.

Senator Collins's top priorities include expanding access to higher education, a goal that led her to coauthor the 1998 Higher Education Act, and continuing her long-standing efforts to help small businesses prosper and create jobs. Among her legislative accomplishments in the 105th Congress was her successful effort, with Senator Richard Durbin of Illinois, to repeal a $50 billion tax break for the tobacco industry.

Senator Collins's civic and professional honors include: Guardian of Small Business Award, from the National Federation of Independent Businesses, for her strong support for small business; 1998 Advocate for Education Award, from the College Board, for her initiatives to make higher education more affordable; Uncommon Service Award, from Maine Common Cause, for her consistent support of campaign finance reform; 1999 Legislator of the Year award, from the Visiting Nurses Association of America; Golden Carrot Award, from Public Voice, in recognition of her leadership on food safety; Friend of the Farm Bureau Award, from the Maine Farm Bureau; and honorary degrees from Husson College and St. Lawrence University.

DIANNE FEINSTEIN

As California's senior senator, Dianne Feinstein has built a reputation as an independent voice, working with both Democrats and Republicans to find commonsense solutions to the problems facing the state and the nation. Reelected on November 7, 2000, to her second full six-year term, Senator Feinstein has served for eight years in the United States Senate. She was first elected in 1992 to fill the remaining two years of Senator Pete Wilson's term when he resigned to become California's governor. In 1994, she was elected to her first full six-year term in the Senate.

Senator Feinstein is the first woman to serve on the Senate Judiciary Committee, where she is the ranking member on the Technology and Terrorism Subcommittee. She also serves on the Senate Appropriations Committee and the Rules and Administration Committee. This year, Senator Feinstein became a member of the Select Committee on Intelligence and the Energy and Natural Resources Committee, where she will work on measures to solve California's electricity crisis.

A leader in the battle against cancer, she cochairs the Senate Cancer Coalition and is vice chair of the National Dialogue on Cancer, with former president George Bush and his wife, Barbara. This is a coalition of more than 130 cancer organizations seeking to improve research, care, and treatment and to find a cure for cancer. Feinstein has raised more than $19 million for breast cancer research through the creation of the Breast Cancer Research Stamp.

In 1994, Senator Feinstein won one of the toughest battles of her career with passage of the Assault Weapons Ban, prohibiting the manufacture and sale of nineteen types of military-style assault weapons.

Other noteworthy legislation sponsored by Senator Feinstein includes the California Desert Protection Act, which protected more than seven million acres of pristine California desert and established the Death Valley and Joshua Tree National Parks and the East Mojave Natural Preserve; the Comprehensive Methamphetamine Control Act, which established new controls over the manufacture of methamphetamine and increased the criminal penalties for its possession and distribution; the Lake Tahoe Restoration Act, which preserves this treasured natural resource; the Gun-Free Schools Act, which sets a zero-tolerance policy to keep America's schools gun-free by requiring all public schools to expel students who carry a gun to school; the Foreign Narcotics Kingpin Designation Act, which enabled the United States to block and seize assets of narcotics traffickers who pose threats to the nation's security, foreign policy and economy; and the Headwaters Forest Agreement, which saved the largest privately held stand of uncut old-growth redwoods. The agreement also helped preserve twelve additional groves of ancient redwood trees and provided strong protections for the endangered marbled murrelet and coho salmon.

Senator Feinstein is a native of San Francisco, and was appointed by Governor Pat Brown to the women's parole board in 1960 at age twenty-seven. In 1969, she was elected to the San Francisco County Board of Supervisors, where she became the first woman president of the board.

She became mayor of San Francisco in November 1978, following the assassination of Mayor George Moscone and Supervisor Harvey Milk and demonstrated a steadiness and command that calmed the city during that turbulent time. The following year she was elected to the first of two four-year terms. As the city's first woman mayor, Dianne Feinstein managed the San Francisco's finances with a firm hand, balancing nine budgets in a row. In 1987, *City and State* magazine named her the nation's "Most Effective Mayor."

Feinstein was born on June 22, 1933, the daughter of a respected surgeon and professor of the University of California at San Francisco Medical School. In 1955, she received a B.A. in history from Stanford University, where she served as student body vice president. She is married to Richard C. Blum, and has one daughter, Katherine; three stepdaughters, Annette, Heidi, and Eileen; two granddaughters, Eileen and Lea; and one grandson, Mitchell.

KAY BAILEY HUTCHISON

Senator Kay Bailey Hutchison, Republican of Texas, is the first woman to represent her state in the U.S. Senate.

Her defense and military construction subcommittee assignments on the key Senate Appropriations Committee give her an excellent opportunity to help shape United States defense policy. In addition, Senator Hutchison serves on the Senate Commerce Committee, where she chairs the Subcommittee on Surface Transportation and Merchant

Marine. As subcommittee chair, she drafted and passed the Ocean Shipping Reform Act of 1998, giving U.S. ships, ports, and shippers a more level playing field to compete internationally. In 1997, Congress passed her Amtrak Reform and Accountability Act, deregulating the railroad and putting it on the path to operational self-sufficiency.

Senator Hutchison spearheaded the effort to pass legislation ensuring that individual states' tobacco settlement funds are protected from seizure by the federal government, as well as a bill strengthening health care benefits for veterans and military retirees. She sponsored and passed the federal antistalking bill, which makes stalking across state lines a crime; the Homemaker IRA legislation, which significantly expanded retirement opportunities for stay-at-home spouses; and the repeal of the marriage tax penalty. The latter measure was vetoed as part of broader tax legislation, but is currently being reevaluated.

Hutchison has emerged as one of the Senate's leading voices on foreign policy and national security issues. She serves as chairman of the Board of Visitors of the U.S. Military Academy at West Point and is a U.S. delegate to the Commission on Security and Cooperation in Europe, also known as the Helsinki Commission. She is cochair of the Congressional Oil and Gas Caucus.

Hutchison's opinion pieces on tax policy, foreign policy, and national security issues have been published in the *New York Times*, the *Wall Street Journal*, the *Washington Post*, and the *Los Angeles Times*, as well as in every Texas daily newspaper.

Hutchison was named "Border Texan of the Year" in 2000; the National Conference of State Legislatures' "Outstanding Member of Congress" in 1998 and 1999; and one of the Texas Women's Chamber of Commerce's "100 Most Influential Texas Women of the Century" in 1999. In 1997, Hutchison was inducted into the Texas Women's Hall of Fame.

She received the Clare Booth Luce Policy Institute Conservative Leadership award in 1999; the Advocate for Education award from the College Board in 1999; the Texan of the Year award from the Texas Legislative Conference in 1997; and was named Republican Woman of the Year by the National Federation of Republican Women in 1995.

Hutchison grew up in La Marque, Texas, and graduated from the University of Texas and University of Texas Law School. She was twice elected to the Texas House of Representatives. In 1990, she was elected Texas state treasurer, and during her tenure, she trimmed her agency's budget more than any other state official while increasing returns on Texas investments to a historic $1 billion annually. She spearheaded the successful fight against a state income tax.

Senator Hutchison lives in Dallas with her husband, Ray, a former colleague from the Texas House. He is a partner in the law firm of Vinson and Elkins.

Hutchison's heritage in Texas is historic. Her great-great-grandfather, Charles S. Taylor, one of Texas's earliest settlers, signed the Texas Declaration of Independence.

MARY L. LANDRIEU

As the oldest of nine siblings and the daughter of a popular political leader, Senator Mary L. Landrieu has had plenty of practice at learning the importance of building consensus and brokering compromise—traits that served her well during her first term in Washington, D.C.

One of Senator Landrieu's first steps after joining the Armed Services Committee in January 1999 was to develop a major compromise that broke a five-year partisan deadlock. The compromise allowed the Senate to move forward with a policy for allowing development of a national missile defense system, saving the policy from what had previously been a sure presidential veto. Her amendment added language that made it clear that the United States will pursue both the development and deployment of a national missile defense system to protect the nation's borders while still continuing negotiations with Russia to reduce nuclear weapons arsenals.

Landrieu's political career began in 1979, when she was elected at age twenty-three to the Louisiana House of Representatives and served on the powerful Appropriations Committee. After two House terms, she served eight years as state treasurer, finding innovative solutions to the state's fiscal problems. As the mother of two young children, Senator Landrieu believes our nation can and must do a better job of balancing our budget and educating children for the global challenges ahead.

During her first two years in the Senate, Senator Landrieu served on the Agriculture Committee, an important panel for Louisiana, given the importance of the agricultural industry to the state. While on the committee, she helped pass a $6 billion federal farm relief bill that provided more than $50 million for Louisiana farmers ravaged by drought.

She is a strong proponent of balancing farmers' needs with the state's health. Senator Landrieu led a bipartisan effort to ensure that the health risks of pesticides are evaluated based on sound science, protecting vital pesticides from bans that could devastate the state's agriculture economy. The Regulatory Fairness and Openness Act would give farmers access to the most effective pesticides while protecting people from potentially harmful chemicals.

Senator Landrieu's assignment on the Armed Services Committee is an important one for Louisiana. The state is home to three major military installations and one of the world's largest shipbuilders. The annual economic impact of military and defense-related contracts on the state is more than $6 billion.

As a member of the Energy and Natural Resources Committee, Senator Landrieu is now leading a bipartisan charge to bring an estimated $300 million a year to Louisiana by giving a larger portion of federal offshore oil and gas drilling revenues to coastal states. The

Reinvestment and Environmental Restoration Act would represent the largest investment in the environment in decades, without raising taxes.

Since the federal government began collecting offshore drilling revenues in 1956, it has taken in more than $120 billion, keeping nearly 100 percent. Under this bill, 50 percent would be redirected to coastal states to preserve coastlines and wetlands. Every state would benefit from additional funding to the Land and Water Conservation Fund and the Wildlife Restoration Fund.

More than 65 percent of new job growth in Louisiana in the past decade was created by small businesses, making them an essential part of the state's economy. Senator Landrieu, who also sits on the Small Business Committee, helped pass legislation that has lessened burdensome federal regulations and created tax relief for small businesses. In fact, her pro-growth, pro-business voting record in the 105th Congress earned her the U.S. Chamber of Commerce's Spirit of Enterprise Award.

While Louisiana has made great strides in many issues in recent years, caring for its children remains a weak point for the state. Statistics on the overall health and well-being of Louisiana's children are among the worst in the nation. Thirty-two percent of the state's children live in poverty, while more than 20 percent of teenage girls give birth before their eighteenth birthday.

In June 1998, Landrieu joined a number of state leaders and child advocates to launch "Steps to Success," an early childhood development initiative aimed at ensuring that all children are ready to start school. The public/private partnership focuses on increasing learning opportunities for children from birth to age three, the period of time during which 90 percent of a person's brain develops.

Senator Landrieu also strongly supports an increased tax credit for families that adopt children with special needs. She would like to see the tax credit for adopting special-needs children increased from $6,000 to $10,000.

BLANCHE L. LINCOLN

First elected to Congress in 1992, Democrat Blanche L. Lincoln is a native of Helena, Arkansas, where the Lambert family still resides. She comes from a seventh-generation Arkansas farm family and is the daughter of retired farmer Jordan Lambert Jr. and Martha Kelly Lambert. Senator Lincoln received a bachelor's degree from Randolph-Macon Women's College in Lynchburg, Virginia, and studied at the University of Arkansas in Fayetteville. Blanche and her husband, Dr. Steve Lincoln, are the proud parents of twin boys, Reece and Bennett Lincoln.

On November 3, 1998, Blanche Lincoln made history when she became the youngest woman ever elected to the United States Senate and only the second woman to win a U.S. Senate seat representing Arkansas since Hattie Caraway in 1932. Lincoln was sworn in as Arkansas's thirty-second U.S. senator on January 6, 1999.

Prior to her tenure in the Senate, Lincoln's election to the House in 1992 marked the first time a woman had won this post representing the First Congressional District of Arkansas. As a U.S. congresswoman, she used her influence to champion children's and women's health issues, as well as to foster rural development. Lincoln's hard work and dedication earned her a seat on the powerful Energy and Commerce Committee, from which she wielded influence on key issues affecting her district. Seven of her bills became law during the 1995–1996 session, when Republicans controlled Congress. Lincoln's announcement that she would not seek reelection in January 1996, due to her pregnancy with twin boys, ended her brief but successful career as a U.S. congresswoman.

Blanche Lincoln brings two important traits to government: a commonsense approach to solving problems and the ability to build a consensus between two political extremes. In the House, Lincoln was among a group of twenty-three fiscally conservative Democrats, called the Blue Dog Coalition, who led the way to balancing the budget. Their groundbreaking fiscal proposal in 1995 was the foundation for the Balanced Budget Act of 1997. Lincoln's ability to facilitate compromise proved beneficial, because she served in a House ruled by Democrats in her first term and by Republicans during her second.

Now in the Senate, Lincoln is a founding member of the Senate New Democrat Coalition, a group of moderate Democrats committed to seeing results from government. Lincoln is also a member of the Agriculture Committee, the Energy and Natural Resources Committee, and the Special Committee on Aging. In addition, she serves on the Senate Social Security Task Force. Lincoln intends to work closely with colleagues from both sides of the aisle to advance issues critical to the future of our nation, including reforming Social Security, keeping our fiscal house in order, improving education, and addressing the agriculture crisis in Rural America. She has already introduced bipartisan proposals on agriculture, education, tax reform, and children's health.

Lincoln serves as a member of numerous service organizations and was recently named one of the Ten Outstanding Young Americans for 1999 by the Junior Chamber of Commerce for her dedication to public service and her efforts to improve the lives of all Americans.

BARBARA A. MIKULSKI

Born on July 20, 1936—the great-granddaughter of Polish immigrants who owned a local bakery—Barbara Ann Mikulski is the oldest of three daughters born to Christine and William Mikulski. She was born and raised in historic and ethnically rich East Baltimore, where her parents ran a neighborhood grocery store across the street from their Highlandtown home. During her high school years, she worked in her parents' grocery store, delivering groceries to seniors in her neighborhood who were unable to leave their homes. Her love for her hometown has never diminished, and throughout her career, she has returned home each night to the city of Baltimore.

After graduating from Mount Saint Agnes College and the University of Maryland with a degree in social work, Barbara Mikulski went to work on the front lines in President John Kennedy's war against poverty. Just as she was considering going back to school, the Baltimore political machine announced a sixteen-lane highway would be built through the historic Fells Point area of Baltimore. This highway not only threatened Fells Point but would have cut through the first black home ownership neighborhood in the city and would have prevented the successful development of the Baltimore Harbor area. She mobilized communities on both sides of the city and stopped the highway, thus becoming known as the street fighter who beat the road. This led to a seat on the Baltimore City Council for five years and then on to the United States House of Representatives in the star-spangled class of 1976.

In 1986, when Senator Mack Mathias decided to retire from the U.S. Senate seat representing Maryland, Barbara Mikulski stepped up to the plate and won an amazing 61 percent of the vote. She was the first Democratic woman to hold a Senate seat not previously held by her husband, the first Democratic woman to serve in both houses of Congress, and the first woman to win a statewide election in Maryland. Mikulski's pioneering efforts and her advocacy on behalf of women candidates have helped elect nine new Democratic women to the United States Senate during her tenure and have made her the unofficial "Dean of Senate Women."

In 1994, Senator Mikulski was unanimously elected as secretary of the Democratic Conference for the 104th Congress, the first woman to be elected to a Democratic leadership position in the Senate. She was elected in 1998 for a third term with 71 percent of the vote and was the first candidate to win over one million votes in the state of Maryland. She is the lead Democrat on the appropriations subcommittee that funds: the U.S. Department of Housing and Urban Development, the Department of Veterans' Affairs, NASA, the EPA, and sixteen other government agencies. She continues to work as a member of the leadership to form the Democratic agenda in the Senate.

Mikulski is recognized as a national leader on the issue of women's health care. In 1992, she assembled a bipartisan coalition to create the Mammography Quality Standards Act, ensuring that whether a woman has a mammogram in Baltimore, Maryland, or Berkeley, California, the standard of quality is the same. As one of the originators of the AmeriCorps concept, she has been called the godmother of national service and a champion for the rights of working people. From potholes to public education, she has a reputation for finding solutions to constituent problems that effect one or one thousand Marylanders. She is proud to be the senator from Maryland and for Maryland. From city activist to United States senator, she has never changed her view that all politics and policy is local, and that her job is to serve the people in their day-to-day needs as well as to prepare this country for the future. She is frequently heard in the halls of the Senate promoting her belief that, "each one of us can make a difference, but together we can bring about change."

PATTY MURRAY

Senator Patty Murray—working mother, educator, local school board president, and state legislator—was first elected in 1992 to represent the interests and values of the people of Washington State. Reelected in 1998 by a wide margin, she is recognized for her commonsense approach and her dedication to the concerns of working families.

Senator Murray is a leader on education and women's issues and serves on the Senate Health, Education, Labor, and Pensions Committee. Her legislation to hire 100,000 new teachers was passed in 1998 as part of the bipartisan budget agreement. She has also written and passed legislation to put computers in classrooms, improve technology training for teachers, and protect children from Internet pornography.

Patty Murray is a tireless advocate for Washington State. As a member of the Senate Appropriations Committee, she has secured needed money to strengthen drug enforcement, put 100,000 more police on the streets, set aside wilderness areas, and eased traffic congestion. She is leading the effort to protect the Hanford Reach, the last free-flowing stretch of the Columbia River. And she is working to increase economic prosperity and stimulate job growth by supporting trade opportunities for local businesses and ensuring open and free access to foreign markets. Murray also serves on the Senate Budget Committee, where she has worked to rein in the deficit and balance the budget while protecting Social Security and Medicare.

Murray is the daughter of a disabled veteran and is proud to have been chosen to be the first woman to serve on the Senate Veterans' Affairs Committee, where she has fought to protect veterans' benefits.

In 1980, Murray served as a parent volunteer for the Shoreline Community Cooperative School. When the state legislature cut funding for parent-child education programs, Murray went to Olympia to lobby against the cuts. One legislator told her she could not make a difference because she was "just a mom in tennis shoes." The grassroots campaign she subsequently organized saved the program, and galvanized her commitment to fight for things in which she believed.

Born in Bothell on October 11, 1950, Patty Murray is one of seven children. She was educated in the public schools of Bothell and received her bachelor of arts degree from Washington State University. She is married to Rob Murray, and they have two grown children, Randy and Sara.

OLYMPIA J. SNOWE

In November 2000, Olympia J. Snowe was reelected with 69 percent of the vote to continue representing Maine in the United States Senate. With her election in 1994, she became only the second woman senator in history to represent Maine, following the late senator Margaret Chase Smith, who served from 1949 to 1973.

Before her election to the Senate, Snowe represented Maine's Second Congressional District in the U.S. House of Representatives for sixteen years. She was only the fourth woman in history to be elected to both houses of Congress and the first woman in American history to serve in both houses of a state legislature and both houses of Congress. When first elected to Congress in 1978, at the age of thirty-one, Snowe was the youngest Republican woman, and the first Greek-American woman, ever elected to Congress. She has won more federal elections in Maine than any other person since World War II.

In the Senate, Snowe has carved out a reputation as a leading moderate, focusing her attention on efforts to build bipartisan consensus on key issues. In 1999, she was cited by *Congressional Quarterly* for her centrist leadership, and is cochair with Senator John Breaux (D-Louisiana) of the Senate Centrist Coalition, a vehicle for communication and cooperation between Senate Democrats and Republicans. In her first term she was appointed to leadership as a deputy whip, and in 1997 was elevated to the position of counsel to the assistant majority leader.

Senator Snowe has worked extensively on a number of issues: the budget; education, including student financial aid and education technology; national security; women's concerns; health care, including prescription drug coverage for Medicare recipients; oceans and fisheries; and campaign finance reform. She has also led efforts important to Maine, including a successful push for federal disaster funds in response to a devastating 1998 ice storm; increased funding for the Togus veterans hospital; reauthorization of the Northeast Dairy Compact, so critical to the survival of Maine's small family dairy farms; and opposition to a proposed federal rule that would have devastated the state's lobster fishery.

In 2001, Snowe became the first Republican woman ever to secure a full-term seat on the Senate Finance Committee, and only the third woman in history to join the panel. The committee is considered one of the most powerful in Congress because its members write tax, trade, health care, welfare, Medicaid, Medicare, and Social Security–related legislation. Snowe chairs the Subcommittee on Health Care, which oversees matters related to health insurance and the uninsured.

A member of the Senate Committee on Commerce, Science, and Transportation, she also chairs its Subcommittee on Oceans and Fisheries, which oversees America's fisheries and the Coast Guard. A member of the Senate Budget Committee, she was a key voice in establishing education as a priority within the context of the first balanced budget since 1969, and in 1999 and 2000 authored the amendment that for the first time created a reserve fund for a Medicare prescription drug benefit. She is also a member of the Senate Small Business Committee.

Prior to her service on the Finance Committee, Senator Snowe had been the fourth woman ever to serve on the Senate Armed Services Committee, where she was the first woman senator to chair the Subcommittee on Seapower, which oversees the Navy and Marine Corps.

Formerly Olympia Jean Bouchles, she was born on February 21, 1947, in Augusta, Maine. She is the daughter of the late George Bouchles, a native of Mytilene, Greece, and the late Georgia Goranites Bouchles, whose parents emigrated to America from Sparta. After the death of her parents, she was raised by her aunt and uncle, Mary and the late James Goranites of Auburn, Maine. She attended St. Basil's Academy, a Greek Orthodox school in Garrison, New York, and graduated from Edward Little High School in Auburn. She earned a degree in political science from the University of Maine in 1969.

Senator Snowe is married to former Maine governor John R. McKernan Jr. She is a member of the Holy Trinity Greek Orthodox Church in Lewiston, Maine.

DEBBIE STABENOW

History was made in 2000, when Debbie Stabenow became the first woman ever elected to the United States Senate from Michigan. From the county commission to the state legislature to Congress, outstanding leadership and hard work have characterized her years in public service.

Whether working to make sure children have access to the best education in public schools, advocating for fiscal responsibility and tax relief for the middle class, fighting to preserve natural resources, reducing the cost of prescription drugs, or preserving Social Security and Medicare, Stabenow is fighting for those things that matter most to Michigan families.

In the Senate, Stabenow's leadership and experience were rewarded with four key Senate committee assignments, including the powerful Budget Committee; Banking, Housing, and Urban Affairs; Agriculture, Nutrition, and Forestry; and the Special Committee on Aging. She also serves as chair of the Women's Senate Network for the Democratic Senatorial Campaign Committee.

Born on April 29, 1950, Senator Stabenow grew up in the small town of Clare, Michigan. She attended Michigan State University, where she received her B.A. in 1972 and M.S.W. in 1975. She worked with youth in the public schools before running for public office.

She was first elected to the Ingham County Board of Commission in 1974 and was the youngest and first woman to chair the board (1977–78). She was then elected to the Michigan House of Representatives, where she served for twelve years (1979–90) and rose in leadership, becoming the first woman to preside over that body. She served in the State Senate for four years (1991–94) and was elected to Congress in 1996, where she served for two terms, representing Michigan's Eighth Congressional District.

Senator Stabenow is an accomplished legislator, having authored more than fifty public acts as a state legislator and having brought such critical issues as the cost of prescription drugs to the forefront of

Congress. She has been nationally recognized with more than sixty awards for her leadership on behalf of families and small businesses. Most recently, the National Committee to Preserve Social Security and Medicare recognized her with its top award and the National Association for Home Care named her a Home Health Hero.

Senator Stabenow's home is in Lansing, where her mother and two children, Todd and Michelle, reside. She is a lifelong United Methodist and member of Grace United Methodist Church.

INDEX